Concrete in Extreme Environments

Edited by
John W. Bull

Eur Ing, BSc, PhD, DSc, CEng, FICE, FIStructE,
FCIHT, FIHT, FIMMM, AFHEA
Professor of Civil Engineering, Northumbria University, UK
and
Xiangming Zhou

BEng, MEng, PhD, PGCert, FHEA, SMRILEM, MICT, MIMS,
MASCE, MACI, GMIStructE, GMICE
Professor in Civil Engineering Design, Brunel University London, UK

WHITTLES PUBLISHING

Published by
Whittles Publishing,
Dunbeath,
Caithness KW6 6EG,
Scotland, UK
www.whittlespublishing.com

© 2018 J. W. Bull & X. Zhou
ISBN 978-184995-327-6

Printed in Malta by Melita Press.

Contents

The Editors

John W. Bull is a Professor of Civil Engineering at Northumbria University, Newcastle upon Tyne in the UK. He graduated from Cardiff University with a first class honours degree in civil and structural engineering and was awarded the Page Prize for the student with the highest marks. His PhD was entitled 'Analysis of Shell Structures by Finite Elements' and his DSc 'Computational Engineering Applied to Engineered Structures' reflecting his substantial contribution to the advancement of knowledge. He has 180 publications including 18 authored or edited books.

Following his PhD, he worked in industry designing and constructing bridges, roads and other civil engineering structures. He has been a professor at universities in Australia and Japan, a specialist assessor in civil engineering for the Higher Education Funding Council for Wales, an EPSRC assessor and a reviewer for South Africa's National Research Foundation. He advises publishers on book proposals, referees papers for journals, has given keynote lectures at conferences and has provided industrial consultancy services. He is also an external examiner for civil engineering degrees. John has been a BSI Committee member for 'Timber Testing Methods', and is a member of the BSI Committee for the 'Structural use of Aluminum'. He has been a member of International Journal Editorial Boards and has been an Editorial Board member for many international conferences.

Xiangming Zhou is a Professor in Civil Engineering at Brunel University London. He received his BEng in civil and structural engineering in 1997 and master of engineering in structural engineering in 2000 both from Tongji University, China and his PhD in civil engineering from Hong Kong University of Science and Technology in 2005. After his PhD study, he conducted post-doctotral research at both Hong Kong University of Science and Technology and Hong Kong Polytechnic University. He became a lecturer in structural engineering at the University of Greenwich in 2007 and moved to Brunel University in 2008. His research interests encompass cement and concrete materials such as advanced concrete technology (e.g. extrusion), low carbon and energy efficient cementitious materials, geopolymer/alkali-activated cement, testing and modelling of rheology and early-age engineering properties of cement-based materials, fracture testing and modelling of hardened concrete, and the restrained shrinkage cracking of concrete. He has a sustained track record of high quality research publications in mainstream construction materials and structural engineering journals. He has led several European Commission framework programme funded research and innovation projects as well as UK research council funded projects. He has conducted peer review of grant proposals for organisations such as EPSRC, European Commission Horizon 2020 Research and Innovation Programme, the Royal Society, and the Hong Kong Croucher Foundation. He is a reviewer for more than 50 international academic journals and supervises PhD and MSc theses.

The Authors

Carmen Andrade researched industrial chemistry and until last year studied the durability of buildings, with particular attention to the corrosion of steel in reinforced concrete at the Institute of Construction Sciences "Eduardo Torroja" of the CSIC, Spain. She is now visiting research professor at the International Centre for Numerical Methods in Engineering (CIMNE). She is the author of numerous papers, has been editor of several books, and has supervised more than 30 PhD theses. She has participated in and led various standardization committees and has been chairperson of several international organizations related to her specialty including UEAtc, RILEM, WFTAO and Liaison Committee. She has received several awards: R. N. Whitney Prize 2013 from the NACE, Robert L'Hermite Medal 1987 from RILEM, and the ALCONPAT Prize for a meritorious career. She has been Director General of Technological Policy for the Ministry of Education and Science, and Advisor to the Secretary of State for Universities in the Ministry of Science and Innovation, Spain.

Emanuele Brunesi is a researcher at the European Centre for Training and Research in Earthquake Engineering – Eucentre. After his undergraduate studies at the University of Pavia, where he received his BSc degree (*magna cum laude*) in civil engineering and his MEng degree (*magna cum laude*) in structural engineering, he started collaborating with Eucentre on national and international research projects for the European Commission, the Italian Department of Civil Protection and private companies. Emanuele received his MSc and PhD in earthquake engineering and engineering seismology from the School for Advanced Studies at Pavia (IUSS Pavia – ROSE Programme). He has been a chartered professional engineer in Italy since 2011. He has authored peer-reviewed journal papers focused on computer methods in applied engineering and mechanics, field observation of failures and experimental testing of structures and structural subassemblies/components. He is a reviewer for several international journals, speaker at international and national conferences and co-advisor for PhD, MSc and MEng theses.

Daniele Cicola graduated in civil engineering with structures specialisation at the University of Pavia, Italy, in April 2013. In 2015 he received the second level masters degree in earthquake engineering and engineering seismology at the Understanding and Managing Extremes Graduate School (UME School), organized by IUSS Pavia (School of Advanced Studies, Pavia). He joined Eucentre (European Centre for Training and Research in Earthquake Engineering) in March 2015 in the structural analysis section. His research interests are focused on seismic behaviour of precast concrete structures and storage tanks.

Giulia Fagà received her masters degree in construction engineering and architecture in 2012 from the University of Pavia. Since 2013, she has worked at Eucentre in the structural analysis department. During 2014–2015 she collaborated on the Prisma project and in 2016 she co-authored the book *Temporary houses, from Paleolithic curtains to 3d printers* and has been also involved in surveys of buildings in the aftermath of the Central Italy Earthquake. In 2017 she became the managing editor of the magazine *Progettazione Sismica* and she is currently the head of the Education Department of Eucentre.

Dr. Dominique Guinot is currently Technical Director for the Waste Water and Technical Concretes market at Imerys Aluminates, France. He holds a PhD in material sciences. After being involved for few years in the development of technical ceramics, he joined the Lafarge Group in 1990 where he has worked in different positions in the field of research and development of cementitious materials, and in technical support of cement plants both in France and Canada. In 2001 in joined Kerneos Aluminates Technology Group (formerly Lafarge Aluminates) as R&D Director and in 2014 became Innovation Director, before reaching his current position.

Dr. Jean Herisson is an R&D engineer at the Imerys Aluminates research centre near Lyon, France. Jean has been working in the field of biodeterioration of cementitious materials in sewer networks for about 10 years. During his PhD, under the direction of Thierry Chaussadent (Ifsttar), he worked on the development of an accelerated and representative biodeterioration test based on the comparison of ordinary Portland cement and calcium aluminate cement mortars. Then he joined the research centre at Kerneos to better understand the specific behaviour of CAC materials in relation to microorganisms. Jean is presently involved in the 4-year French project DURANET, which aims to improve the sustainability of sewer networks through the design of new materials, validated by accelerated laboratory tests and on-site exposures. Jean is also involved in the RILEM technical committee TC-253 MCI with the aim to produce a state of the art report on the deterioration of cement-based materials by microorganisms in different contexts. Finally, Jean is involved in European and French standardization groups to respectively develop and review standards in relation to H_2S corrosion.

Mads Hovgaard is a structural engineer with structural engineering consultancy Rambøll Denmark. Holding a PhD degree from Aarhus University, Denmark in structural engineering, Mr. Hovgaard is specialized in the design and reassessment of advanced concrete structures. He has a keen research interest in probabilistic analysis and machine learning for structural optimization and risk minimization.

Dr Ted Kay is an independent advisor on matters related to concrete production and concrete construction. He has worked for British consultants and also worked as an Advisory Engineer for the Concrete Society. He has worked in the Middle East and published papers on Middle East concrete over an extended period. He is a past President of the Concrete Society and an Honorary Fellow of the Institute of Concrete Technology.

Maurice Levitt is a mathematical physicist who has over 50 years' experience in the construction industry applying physics material science to many problems. He has dozens of publications and three books in his name. He specialized in all aspects of durability and founded the ISA test standardized and specified for both precast and in situ concretes. He still practises consultancy and keeps his knowledge honed to a fine degree.

Neil Milestone gained his MSc from Victoria University of Wellington and completed a PhD in soil chemistry at the University of Waikato. In 1973, he joined the Cement and Concrete Section of DSIR NZ, where he conducted research on the complex nature of cement hydration. Following a sabbatical at the University of Illinois in 1976/77, he was a key member of the DSIR group investigating zeolites prior to construction of the synthetic gas-to-gasoline plant at Motonui. With the growth in geothermal electricity production in the 1980s, Neil went to Brookhaven National Laboratory to study geothermal cementing where he discovered the role CO_2 plays in cement and casing corrosion. When Industrial Research Ltd was formed in 1992, he was appointed Team Leader of the Advanced Inorganic Materials group investigating inorganic coatings, zeolites and catalysts and cement. He has been awarded several industrial prizes and a DSc for his research.

In 2002, Neil was appointed senior lecturer in the Immobilisation Science Laboratory at the University of Sheffield where he rose to be acting director. He returned to New Zealand in 2009, where he is a distinguished researcher at Callaghan Innovation; he is also a visiting professor at UCL, and an adjunct professor at Victoria University of Wellington.

Professor Peter Robery FREng is a chartered civil engineer, founder/director of Robery Forensic Engineering Ltd and visiting professor in forensic engineering at the University of Birmingham. He has worked for consulting engineers and contractors on projects in many parts of the world, specializing in deterioration, maintenance and repair. A past President of the Concrete Society, he is a Fellow of the Royal Academy of Engineering, Institution of Civil Engineers and Institute of Concrete Technology.

Dr. François Saucier is a civil engineer working as senior business unit director at Imerys Aluminates for the Waste Water and Technical Concretes market. His Ph.D. on the durability of the bonding of concrete repairs is from Laval University, Quebec City, Canada. He won the 1992 ACI Wason Medal for materials research. After founding a concrete consulting company in Canada, he joined Lafarge in France in 1996 and has worked since then in different positions related to concrete technology. Since 2008 he has been in charge of the Waste Water market and has developed a broad knowledge of this area through interactions and experiences with users around the world.

Professor Lupin Tang received his PhD in 1996 at Chalmers University of Technology, Gothenburg, Sweden. Since then he has worked at SP Technical Research Institute of Sweden for 12 years and rejoined Chalmers in 2008 as professor and research leader for building materials. His main research interest is new types of cementitious materials and durability of concrete, especially chloride transport mechanisms and chloride-induced corrosion of steel in concrete. He has over twenty five years' experience in the field of electrochemical applications in construction materials.

Don Wimpenny is a chartered materials engineer at CH2M Hill (now Jacobs). His work includes the design and specification of new concrete structures, as well as investigating existing facilities and construction material failures. He was worked on the durability design of numerous projects around the world. He is a Fellow of the Institute of Materials, Minerals and Mining, the Institute of Concrete Technology and the Concrete Society.

Yan Xiang is a professor of engineering as well as the deputy director at the department of dam safety management of Nanjing Hydraulic Research Institute, China. During this time, he has chaired and participated in national scientific research projects, such as the National Key R&D Plan, National Science and technology support program project, special major projects for water conservancy and public welfare industry, and three national natural science funds. His current research interests include safety assessment of water conservancy and hydropower projects, numerical simulation of hydraulic structures, and dam safety monitoring. He has directed and completed the safety assessment of many important reservoirs in China such as Xiao Langdi reservoirs, Xi Xiayuan reservoirs, Si Minghu reservoirs and participated in the compilation of more than three national industry specifications.

Chapter 1
Introduction

Maurice Levitt

1.1 Background

For readers with an interest in the design, manufacture and use of concrete in extreme environments, the following chapters will be very useful as they recount the individual authors' actual experiences on the use of concrete at its limits. The 'Trekkies' among you should not assume that you are necessarily venturing where no-one has been before, but use existing experience wherever possible. Coupling a thoughtful extension and applying engineering and science to address extreme demands is the logical way forward. At the same time, it needs to be borne in mind that there are likely to be some environments where, no matter how deeply investigated and researched, concrete should not be used. The analogy with limit state design in structural concrete springs to mind.

Concrete is a multi-variable material, and its performance criteria depend on all variables, some having more importance than others. These variables form links in a continuous chain joining the starting design point to the on-site end performance point. As stated in the preceding paragraph, the first link in tackling extreme environments is to address these demands by thinking very deeply about which links in the production chain will require more emphasis than one would typically apply. The main danger probably lies where the design team does not take into account that the expertise of other disciplines may well be required and so they mistakenly adjust their existing parameters on a predictive rather than a logical basis. In the following chapters' experiences it will be seen that this need was appreciated and actioned accordingly.

The past lack of multidisciplines can be exemplified by the problems experienced in both *in situ* and precast concrete in the period from 1945 to *c.*1980 in terms of the engineering approach to design and use. A little bit of chemistry was added in the 1950s when it was found that one of the ingredients in Portland cement was more prone to sulfate (back then it was spelt 'sulphate') attack than the other cement ingredients, so a cement was marketed with a reduced amount of this ingredient and sold as 'sulfate-resisting cement'. This led to its use in sulfate risk applications, but with less care given to the other variables, especially water content. The name of the cement gave false confidence in the performance of the concrete. Similar examples of culpability in concrete's history are not rare and there has inevitably been a blame philosophy.

When one uses the word 'extreme' it means that one is seeking to use or explore the application of concrete outside the spectrum of known variables. This, in turn, means one is considering using above the highest or below the lowest end of past experience. Whatever the reader's profession it cannot be emphasised too strongly that preparation for an extreme environmental application needs a team input and not that of a single professional. Examples are given – both related and unrelated to extreme applications from the authors' experience – illustrating that one has to encompass all the aspects of the environmental effects to succeed in tackling the problems that will arise.

In this respect, experience and logic play important roles and the following chapters give the reader an insight into actual applications of concretes deployed in extreme environments. Notwithstanding this, one is likely to encounter other extreme environment scenarios, and whether or not the application is classified (as distinct from an open contract), every relevant avenue needs to be explored, no matter what obstacles are met.

In reference to laboratory pre-contract tests simulating the extreme environment in its effect on various concretes, beware of the attraction to make the test environment even more extreme than that envisaged with the aim of accelerating the effects. Great care needs to be exercised in any such assessment because, sometimes, the accelerating conditions can render the effects less severe, possibly due to locking or buffering conditions. It would be more logical to test under as identical a condition as possible for an application in the future so that simulated experience data can be stored for future reference. Many years ago the then Building Research Station (now BRE, The Building Research Establishment) used to give in their talks a 'rule of thumb' for products and systems on their natural weathering site, which could possibly be applied to a simulated environmental test on concrete:

If it performs well for 1 year it will last for 2 years.
If it performs well for 2 years it will last for 5 years.
If it performs well for 3 years it will last for 10 years.
If it performs well for 5 years it will last for 20 years.

This discussion indicates that, in addition to current demands, one should prepare for future ones. It is relatively simple to predict future extreme environmental uses, but the cost and control of any simulation or assessment would need to be underwritten. In reading this and the following chapters, the seed may well be sown such that leading organisations in the UK or other countries may take on board the logic of preparing for the future rather than being saddled with a possibly dangerous urgent requirement.

The author of this chapter is a mathematical physicist by profession, with six decades of professional work as a construction materials and systems consultant, researcher, trainer and demonstrator. As one can imagine, considerable experience has been acquired and one's own mistakes as well as those of others will be found germane in illustrating where one's own experience or that of others shows a shortfall. Although the examples in the following subsections do not all relate to extreme environments, the lessons learnt show the relevancy of 1) not thinking the whole thing through and 2) not engaging the opinion of other disciplines. In order to preserve some degree of anonymity, the names of people and organisations have sometimes been deleted, but all examples are factual. The reader may well identify lessons to learn from the ensuing examples and apply these to future challenges they may face.

Whether or not the environment is extreme or some other factor is outside its normal range of use is not necessarily the issue. There is generally always something to learn from anything one encounters. It is tantamount to (author's opinion) quoting the saying that 'Travel broadens the mind' when it is only true to say 'Any experience broadens the type of mind that can be broadened'. So see what you can glean from the following author experiences.

1.2 The chromate ion

Although subject to the Official Secrets Act (as I still am) when I was working at the Atomic Energy Research Establishment at Harwell, this particular item was declassified as it ended up in an Open Day, with the then Minister for Works and many other notables and the press present at its demonstration. The basis of the research was an attempt to take chromate ions out of effluent, and it was found that a special blue-coloured almost transparent water spider (measuring about 100 microns across) would eat nothing but chromate ions. These ions are not only poisonous to marine and river life and flora, but are poisonous and carcinogenic to humans, among other risks.

The efficacy of the research was demonstrated in the laboratory with the spiders in an aquarium tank holding about 20 litres of water, giving the tank an attractive blue colour. A technician slowly poured a yellow-coloured 500 ml beaker of chromate solution into the tank and the water turned green. After a short interval of spiders gorging on chromate, the contents returned to the original blue colour. Now to the Open Day.

A small 2-m-deep swimming-pool-sized tank measuring about 20 m × 10 m was filled with spiders to the same concentration as the laboratory tank, giving it the same attractive blue colour. With about 30 people present, a technician opened up a 10 cm valve and quickly let in the same ratio of chromate solution as used in the laboratory assessment. The critical words in this and the preceding paragraph are 'quickly' and 'slowly' as, much to the horror of the spectators, the water turned a muddy black/grey as the influx was too fast for the spiders to consume the chromate and they were all poisoned and died. The main lesson learnt from this was about the size and speed effects moving from a model to a full-sized application. This 'size' effect leads us to the next example.

1.3 Micro-concrete

The author used to work for the then named 'The Research Committee for the Cast Stone and Cast Concrete Products Industry', a name, incidentally, that was inversely proportional in length to the number of people it employed! It was based at Wexham Springs, working closely alongside the Cement & Concrete Association (C&CA) so they were aware of each others' work. Around 1960–1965 the C&CA investigated micro models of structures, scaling down to about 1:50.

In order to simulate a concrete structure with a model, a rich sand/cement mortar was used for sections. After curing and maturing, the model was loaded and strain gauge readings taken. The data were applied to a full-scale model and found to be in serious error, because the load/strain relationship predicted from the model did not take into account the different behaviour of a mortar mix compared to concrete with a 20 mm maximum size aggregate.

One lesson learnt from this is to be careful of the size effect when changing from scaled-down models to a full-sized structure. As a physicist familiar with Hooke's Law, I queried the calculations, as the designers had assumed that the linear part of the stress/strain curve in compression could be extrapolated through the origin into the tensile zone in a continuous straight line. No satisfactory answer was received.

1.4 The PVA admixture debacle

In the early precast concrete research years I was instructed to investigate American research on the addition of polyvinyl acetate (PVA) emulsion to concrete, as a leading US/UK company had launched the product onto the UK market with claims (as recollected) of a 40% improvement in compressive strength, 30% in bending strength and many other alleged positives. The company was more than pleased to supply our laboratory with samples, and concrete cubes and prisms were made to examine the effects. When tested at 28 days old, the above percentages were obtained but with negative signs before each.

The US origins of the claims were investigated and it was ascertained that three Ivy League University postgraduates had obtained their PhD awards by undertaking this work. Access to their theses was obtained and it was found that their reports were correct but all their research had been undertaken on 12.5 mm mortar cubes. A link was set up with a polymer chemist to investigate why the performance in concrete was opposite to that in mortar and, without going into the science too deeply, it was found that polymers form continuous reinforcing 'skins' when they lose the water in the emulsions.

A tiny mortar cube will lose this water quickly because it has a large specific surface compared to a 100 mm or 150 mm concrete cube and polymer *in situ* film formation will result. On reporting the findings to the UK branch of the large US/UK chemical company, the product was withdrawn from the market and purchasers were reimbursed and deliveries returned.

1.5 Flag and kerb paper usage in manufacture

Further ignorance of deeper chemistry was exemplified in a research project where paper was used in the manufacture of hydraulically pressed products. Here, a concrete mix with a water content on the order of twice that used in structural concrete was poured into a press, with the base lined with a sheet of paper, allowing excess water egress under hydraulic pressure. Although modern production often removes the paper by multi-jet blowing, at the time of the research it was removed manually when the products were 24–48 hours old. Sometimes this resulted in clean removal, but at other times the paper stuck partially or fully to the concrete. The research project aim was to alleviate or annul the sticking problem.

With the then knowledge of chemicals that would retard hydration without any known harmful effects, sample sheets of the customary brown paper were treated with zinc sulfate solution and allowed to dry. They were then distributed to three manufacturers, who were asked to use them and report back on their performance. The effect was exactly the opposite of the intention, with one manufacturer suggesting that we advise one of the UK leading wallpaper manufacturers of the treatment. There was no way that the treated paper would strip from the concrete.

The reason for the sticking became obvious, and it was not because the zinc retarded the setting time, but because the sulfate reacted with the free lime released during the hydration to form a virtually insoluble calcium sulfate. This locked the paper to the surface of the concrete. All this roused curiosity, and a visit was made to a local paper manufacturer, albeit one specialising in tissue production. The pulp was fed onto the common Fourdrinier processor after it had been fibre-processed and alum size (potassium aluminium sulfate) added. This immediately 'rang a bell' and, after sending an apology to the cooperating flag manufacturing companies, a request was made for each company to sample six sheets from each batch being used and to send the 'before', 'during sticking' and 'after' sheets back to our laboratory for analyses.

All the results showed a peak in sulfate for the papers used in that production at the 'during' phase, and this resulted in a request to the paper manufacturers to supply paper with alum size sulfate below a specified level. This resulted in the problem being alleviated. The research also led to some manufacturers using pure white low size paper, which, although about twice the price per ream of the usual brown paper was half the weight, so the thinner sheets worked out at the same sheet price.

The lesson learned from this episode was that all aspects of a chemical reaction needed to be studied as well as looking for problem avoidance by substitution.

1.6 The hardness/strength relationship

In the 1950s and early 1960s, before the introduction of the non-destructive testing (NDT) rebound hammer, manufacturers as well as research units sought NDT tools that would give an indication of, or a direct relationship to, the compressive strength of the concrete. All members of the aforementioned research committee were directors of precast concrete companies, with the chairman running a structural company manufacturing columns, beams and similar units using vibration compaction. This company used the Brinell (not a misspelling of Brunel) Hardness Pistol to assess the strength of their units, as they had found that the measured averaged diameter of the indentations was a constant X mm for their averaged cube strengths of Y.Mpa. The Research Committee team was instructed to examine the performance of the Pistol for a range of concrete strengths.

Concrete 150 mm cubes were made with a range of 28 day characteristic strengths varying from 20 to 50 MPa and cured in water tanks controlled at 18–22°C. At 27 days old they were removed from their tanks and allowed to dry until the following day, when each cube was hardness-tested in duplicate on each moulded face. The 20 diameters obtained were averaged, and the cubes were crushed. The design strengths were obtained for each concrete but the average diameter was always the same.

This was not surprising on reflection, as the Pistol was primarily made for testing metals and alloys, but what was probably more relevant was the dissipation of energy. Each time the Pistol was fired, the energy impacted to the surface was constant; a weak concrete absorbs this impact more easily than a strong concrete and this would result in similar indentations for all concretes. The research project was terminated and the Pistol retained for possible later metallurgical use.

1.7 Precision tunnel segments

As it unfolds, this application will be seen to be not unrelated to that described above in Section 1.6. There were two main problems, one in manufacture and the other in the application on site. The first was a dimensional specification never encountered before and the second was the impact on site. The units concerned were curved truncated units of unreinforced concrete, each measuring approximately 300 mm and 200 mm along the longer and shorter arcs, respectively. These dimensions were specified to be within ±0.1 mm. The products would be in production for over a year and the factory, although mainly enclosed, would be subject to temperature differences varying from 10°C to 40°C. Consideration was given to using steel moulds, but a temperature difference of 30 K over the 300 mm would result in a movement of 0.1 mm.

It was found impossible to find a steel mould manufacturer able to supply within the specified tolerance, so the precast concrete manufacturer manufactured concrete moulds and these were allowed to mature then ground down to accept a precision Invar-framed template. Although these moulds were very heavy and had their own built-in lifting devices, their attraction was their temperature inertia, resulting in size and tolerance control.

The units were manufactured many years ago before the advent of fibre reinforcement. They were designed to form the wall of a concrete tunnel approximately 3 m in diameter, with a lower outer half and upper inner half supporting frames. Units were placed hydraulically, alternating the tapers so that the final insertion was a locking unit placed narrower end first into the gap and pressed into position with the hydraulic ram shock-vibrating. This unit was found to suffer from spalling and cracking damage and resulted in a large percentage of scrap. To overcome this defect the precast concrete manufacturer was asked to alter the specification and increase the 28 day cube strength design from 40 MPa to 45 MPa.

When these were used on site, the locking unit always spalled and cracked, with 100% scrappage. When asked to troubleshoot the problem, an examination was made of the Hookes Law curve for concretes of different strengths, and it was found that a weaker concrete was capable of absorbing more impact energy than a strong concrete. As an experimental trial the precast manufacturer made further units with a designed 28 day strength of 35 MPa. This alleviated the site problem almost completely, so the specification was altered thereafter. It had to be borne in mind that a lower strength might have been more attractive, but durability aspects such as sulfate attack and attrition would have probably suffered.

One can observe the similarities in cause and effect in these last two examples, but this latest example also shows that attention must be paid to other properties relevant to performance.

1.8 Rocket exhaust concrete enclosure

In the testing of rocket-fired engines, preliminary testing found that even when using High Alumina Cement with blastfurnace slag aggregate the concrete degraded quickly. The exhaust temperature exceeded 1,500°C and was the cause of the degradation, with the 'weakest links' in the concrete being the alumino-ferrous component of the cement and the fusion into glass of the slag aggregate. Laboratory testing in an electron arc furnace indicated that changing from High Alumina Cement to pure Alumina Cement coupled with the use of alumina aggregate gave vastly superior properties, so it was decided to specify these two materials only for the manufacture of duct-type sections in precast concrete.

Both the cement and the aggregate were 50–100 times more expensive than High Alumina Cement and blastfurnace slag aggregate, so the precast manufacturer hired weighing machines with an accuracy better than 1% and purchased a new pan-type mixer in order to achieve the best mixing possible. From density figures obtained from the simulation experiment, mixes were weigh-batched from materials kept in secure holds. The matured units were transported to the test site and tested under full-scale conditions and were found to suffer only slight surface dusting at the end of each test.

There is no published data on the use of pure alumina cement and aggregate, but this does not mean that it has not been used in an effective refractory application. The dearth of knowledge could well be due to applications being classified, so no-one generally gets to hear or read about them. The classification of applications is reflected in the next example.

1.9 Explosive-proof cladding and roofing

At the time of the 'troubles' in Northern Ireland a precast concrete manufacturer accepted a contract for the production of roof and wall cladding units specified as being 'resistant to sudden and large rapid changes in air pressure'. This was the ambiguous way of avoiding any reference to explosives or terrorist activities. The units were manufactured to a characteristic strength of 50 MPa using granite coarse and natural sand fine aggregates reinforced with 50-mm-long steel fibres.

Trial mixes revealed two problems. First, the fibres had a detrimental effect on workability, and a high workability admixture had to be used. (The contract pre-dated the use of plasticisers.) Second, even with thick gloves, handling the fresh concrete as well as the test specimens caused protruding fibres to pierce gloves and penetrate to the operatives' hands, resulting in injury. This problem was solved by using armour-faced gloves, which, although awkward in gripping and handling, prevented injury. These gloves had to be stored in a warm, dry ventilated area overnight, otherwise their armour started rusting, leading to rust marks being transferred to the concrete.

The client sought information from tests on these concrete units when fixed into this windowless central gas control pumping station, but none was forthcoming. Although the heading 'explosive-proof' has been used, it needs to be borne in mind that the source of possible damage would be conventional explosive chemicals. Proof against extreme blasts such as nuclear or penetrating bombs' air pressure might well be a separate consideration. It is also worth making a comment on the strength specification of 50 MPa in light of the earlier example of the tunnel segments and the Brinell Hardness Tester, as weaker concrete would be more effective in absorbing impact energy.

Purely as a passing comment, it is probable that, as with so many other properties of concrete, one should be thinking about a range rather than a maximum or minimum level of a property. For a concrete used in this extreme application of explosion resistance, perhaps one needs to design for sound resistance, thermal inertia aims and structural requirements. This would possibly lead to designing a concrete with a mixture of lightweight and natural aggregate and a design strength in the range of 30–40 MPa. Assessing the performance of the concrete as far as the current state of knowledge is concerned is not amenable to a simulation program. The other possible disadvantage is that an explosive test assessment would need to be under the control of a military or government activity and would need careful pre-contract discussion.

1.10 Silage

The phrase 'acid attack' in relation to concrete often initiates thoughts of sulfate attack in the mind of a construction person, but sulfate, especially in the form of magnesium sulfate or sodium or potassium sulfate, attacks by ion exchange and not as a dissolution process. For Portland cement-based products this reaction is with calcium salts, which can cause degradation by the formation of calcium sulfate. Another source of acid attack is an industrial process, where the aggressive agent could be inorganic or organic acid, or a mixture thereof.

Organic acids can be quite aggressive, and one extreme environment to which concrete is subjected is when precast or *in situ* concrete is exposed to acids generated in the decomposition of grass to produce silage cattle feed. Grass and its silage product are commonly stored in silo towers, with the mown material blown into the top of the tower with, in dry hot weather, the addition of a water spray to inhibit dusting. In the rotting process, butyric, oxalic and acetic acids are formed, and these severely attack Portland cement concrete. The early philosophy with tower construction products was that degradation would be rapid and the tower would have to be replaced after a few years.

Construction was usually carried out with precast hydraulically pressed paving slabs externally prestressed with external steel cables. The door at the base from which silage was withdrawn for cattle feed had to be a special vibrated precast unit, which also had a short lifetime. It was found that all precast units had to be made to a higher than normal specification to promote a more acceptable lifetime. The stressing steel cables also had to be modified to increase their organic acid resistance.

Another product generated during the rotting of grass is heat, because the reaction is exothermic. It has been known for fires to burn inside the towers and these can last for months without being detected as they are surrounded by cold damp silage.

A couple of observations may be of interest to the reader. First, a north European patent was filed in c.1960 for the 'Ocrat Process', where concrete in its mature state was subject to treatment with hydrofluoric acid gas. The calcium fluoride formed on the surface made a relatively impermeable layer resistant to a large range of both inorganic and organic acids. However, as one may well imagine, the process did not catch on as the capital cost of setting up the system was prohibitive. On top of this, hydrogen fluoride (HF) gas is expensive to produce and attacks almost everything with the exception of common plastics such as polythene.

Like many experiences in life, this promotes a second observation. If HF gas forms an acid-resistant impermeable layer on concrete, are there any other less dangerous chemical approaches that could form a similar or equivalent protective layer? One answer may be anhydrous citric acid, which has been observed in laboratory trials to cause an initially violent reaction when applied to concrete, but this reaction ceases after a few seconds. This may be due to the formation of an impermeable layer of calcium citrate and could possibly be the basis for a research project.

Although these nine examples are not all extreme environment cases the reader will hopefully have gained two main advantages in reading them:

- First, the value of a multi-disciplinary approach is more often than not a necessity in tackling construction problems.
- Second, any experience gained generally always has one or more facets of knowledge that can be added to one's data bank of experience and extracted and applied to other challenges

Therefore, in reading of the experiences of other authors in the following chapters, please keep your mind open, as there is often an indicator that one could find useful in other scenarios. The editors of this book encourage anyone who obtains future experience in the use of concrete in extreme environments to publish their experiences in any one of the prominent technical/scientific journals. Coupled with this it would be appreciated if the person or persons concerned could obtain the formal permission of that journal's publisher (with due acknowledgement) to publish the said article enhanced and updated as a stand-alone chapter for a further book. All this will enable readers to have a single reference on the subject.

Chapter 2
Recognising severe environments

Ted Kay, Peter Robery and Don Wimpenny

2.1 Introduction

Most design codes or industry specifications for concrete will attempt to define the severity of the service environment, typically as exposure classes (British Standards Institution, 2013) or sometimes as more complex numerical rating systems (Alexander *et al.*, 1994).

Away from the more extreme environments, the severity may not be obvious. It can depend on interactions between the concrete's constituents, the service environment, design life and performance requirements. For example, a concrete for a warehouse floor could be subject to abrasion by hard-wheeled forklift trucks, as well as having to meet surface regularity requirements for operating in very narrow aisles. This situation would be entirely different from a floor for a clean manufacturing facility, where the generation of dust or static electricity may need to be controlled, or an imprinted pigmented slab for an architectural application where control of cracking and long-term stability of the colour and finish would predominate.

The engineer should not blindly rely on the codes and published guidance, but instead should carefully consider the specific combination of conditions and requirements of the design.

One way of defining the severity of an environment for concrete could be past performance or, more accurately, inadequate performance, in the laboratory or in service.

This chapter provides eight case studies (A–H) of inadequate concrete performance across a wide variety of environments. Each case study sheds light on the response of concrete elements to chemical, physical and biological processes:

A. Hot saline environments: the coastal zone of the Arabian Peninsula
B. Softwater leaching in a service reservoir
C. Thaumasite sulfate attack to bridge foundations
D. Acid attack to a bund at a water treatment works
E. Fire: the Buncefield incident
F. Abrasion in aggregate storage bins
G. Algae: experiences from Blackpool
H. Mould growth on long-span bridges: a visual problem.

Failure can eventually bring about a revolutionary change in materials as well as the development of codes of practice to address the problem, such as the response to high alumina cement conversion in structural concrete (Bate, 1985). More often, the construction industry undergoes evolutionary change through a sequence of small steps, as further information comes to light and new materials, such as cement types, are introduced into the industry. An example of this is the guidance for concrete in aggressive ground, which was developed based on laboratory and field exposure work undertaken at the Building Research Station (now Building Research Establishment). A series of guidance documents, including BRE Digest 363 (Building Research Establishment, 1991), seemed to provide a sound basis for the specification of concrete in acid and sulfate ground conditions. However, concerns were raised from the late 1980s onwards about the possibility of a different type of attack occurring at lower ambient temperatures. This concern was based on laboratory studies and a small number of field examples. This changed in 1998, when serious deterioration was uncovered in the foundations of an overbridge for the M5 motorway. Thaumasite sulfate attack to major structures was confirmed to be more than a theoretical possibility. Case study C gives some further details of the problems experienced and the action taken.

There are some common traits to these case studies, which should be considered by the designer, constructor or owner of an asset:

- Concrete sometimes needs protection, and failure to inspect and maintain that protection can have severe consequences.
- The service environment and its severity may only become established or apparent once a facility is in service.
- Novel materials and practices may bring benefits, but they can also introduce unforeseen problems.
- A successful solution for one environment may not work as well elsewhere, particularly if there are changes in a number of influential factors, like temperature.

2.2 Case studies

Case Study A: Hot saline environments – the coastal zone of the Arabian Peninsula

Background

The coastal strips of the States of the Arabian Peninsula represent one of the most extreme environments for concrete and arguably constitute the most severe natural exposure anywhere on the planet. This is due to a combination of factors, including high temperature and highly saline groundwater allied to a water table that is close to the ground surface. The high temperatures not only make the production of concrete difficult but also result in a faster rate of reaction if deterioration processes start.

There had been little concrete construction in this geographical location prior to the 1970s. In the mid-1970s a substantial rise in oil prices meant that there were huge sums of money available for the construction of bigger buildings for domestic, commercial and civic use and for the development of modern infrastructures. Airports, highways with bridges and tunnels, power stations, desalination plants and sewage treatment works were constructed within a short time period in most if not all of the states in the region. Ports were extended and some completely new ports were developed.

Many of the projects constructed in reinforced concrete during this period suffered early deterioration problems, sometimes related to sulfate attack on underground elements, but mainly resulting from corrosion of reinforcement. Although there were no known structural collapses, there were a few cases where concrete structures had to be demolished and rebuilt. There were many, many more instances where partial demolition and reconstruction occurred, where structures were subject to localised repair or where additional protection (e.g. cathodic protection) had to be provided. This period in this location can be viewed as a massive 'failure' of the design and construction industries as, in a large number of cases, the industries were incapable of providing concrete structures that had the required durability.

Contributing factors

This failure needs to be reviewed in the context of its time and place. The indigenous design, construction and materials industries could not cope with the increased construction load imposed upon them. The majority of the projects were of a scale and value many times greater than anything that they had tackled previously. Cement and reinforcement were, for the most part, imported. Demand was such that quality was suspect at times. Imported materials were held up offshore for months because of port congestion. Once unloaded, materials were stored inappropriately, with both reinforcement and bagged cement being left in the open, on the ground.

There were some major rock quarries with good quality output, but the majority of aggregate production operations were small-scale borrow pits for sand and gravel, with rudimentary processing and very little quality control. Aggregates were also stored directly on the ground where they could become contaminated by chlorides and sulfates from the soil and groundwater. There were few ready-mixed concrete plants, and batching plants were set up on individual construction sites.

The labour force required to undertake construction was brought in mainly from the Indian subcontinent. Demand was so great that much of this labour force was unskilled, lacking in knowledge of modern construction techniques, and did not have any sort of previous large-scale construction expertise.

Large foreign companies, both contractors and consultants, were employed to cover the shortfall in capability of the indigenous construction industry. This was at a time when the use of chloride as an admixture in concrete had only recently been banned by British Standards and was permitted in American Standards. The ability of chlorides to penetrate concrete from an external source was not widely appreciated. It was not until the amendment to British Standard CP 110 (British Standards Institution, 1972) in May 1977 that strict limits on the chloride content of fresh concrete were introduced (Page and Page, 2007) in Britain. Deterioration of reinforced concrete structures had not been experienced by many engineers in Europe and North America, and concrete was generally viewed as an extremely durable material. The design standards of the time placed little, if any, emphasis on durability. The expatriate engineers who arrived to work in the region used their own familiar domestic design standards (there were no local standards), unaware of the potential consequences in the much harsher environment to which their concrete structures would be exposed.

CP 110

The British Standard for the design of concrete structures, CP 110 (British Standards Institution, 1972), had been introduced only a few years previously in 1972. This standard included some durability requirements for concrete elements both above and below ground in terms of cover to reinforcement in one table and strength grade or cement content and water cement ratio in a separate table. There were five 'conditions of exposure' for the purposes of durability design, as shown in Table 2.1. Curiously, the exposure conditions for cover include a 'Very severe' category, which is not included in the table for cement content and water cement ratio requirements. The requirements for concrete subject to de-icing salts are the same as those for the

Table 2.1 Extract from durability provisions of CP 110 (courtesy of British Standards Institution, 1972).

Condition of exposure	Min cement content (20mm aggregate) kg/m³	When water cement ratio strictly controlled		Min cover (40N/mm² concrete) mm
		w/c ratio	Min cement content (20 mm aggregate) kg/m³	
Mild: e.g. completely protected against weather, or aggressive conditions, except for a brief period of exposure to normal weather conditions during construction.	250	0.65	230	15
Moderate: e.g. sheltered from severe rain and against freezing whilst saturated with water. Buried concrete and concrete continually under water.	290	0.55	260	25
Severe: e.g. exposed to driving rain, alternate wetting and drying and freezing whilst wet. Subject to heavy condensation or corrosive fumes.	360	0.45	330	30
Very severe[1]: e.g. exposed to sea water or moorland water with abrasion.				60
Subject to salt used for de-icing	290	0.55	260	40[2]

[1] This exposure condition appears only in the CP 110 table for cover and does not appear in the table giving minimum cement contents and maximum water cement ratios.
[2] Only applicable if concrete has entrained air.

13

'Moderate' exposure condition except for cover, which admittedly has a big influence on durability. There were not the same concerns at that time about exposure to chlorides as there would be today.

Concrete specifications based on CP 110 were commonly used for construction in the Middle East in the mid to late 1970s, and the engineers who produced the specifications and used them in construction had the reasonable expectation that they would provide durable structures. Durability of concrete structures had not been a concern, within their experience. Sadly, in many cases, this expectation was not fulfilled in practice and there were many durability failures.

Hot weather and other effects

The climate and physical environment of the region surrounding the Arabian Peninsula are extremely challenging. One consideration, sometimes overlooked, is the human physiological aspect. Temperatures, both day and night, are extremely high and these, in combination with high relative humidity, have a debilitating effect that can reduce both physical and mental capacity.

The high temperatures also affect concrete production at all stages. Some of the difficulties associated with high temperature are shown in Table 2.2.

The problems created by high temperatures were exacerbated by the saline environmental conditions. Much of the development and construction took place in the coastal strip, so structures were subject to a marine environment. Both the Red Sea and the Gulf are enclosed, with only narrow connections to the Indian Ocean. The high temperatures and consequent evaporation have led to elevated salinities in these waters. In the coastal region there is also saline groundwater that is close to, and in some cases at, the natural ground surface. There is a tendency to upward leaching in these areas. The fine-grained soils lead to a capillary rise zone that can reach the ground surface. Intense evaporation then deposits high concentrations of salts (chlorides and sulfates), which can be very much higher than those found in the groundwater. These salts are blown around by the wind and are deposited on horizontal surfaces. Early morning dewfall provides the moisture necessary to enable the salts to penetrate

Table 2.2 Effect of high temperature at different stages of concreting.	
Mixing	Increased water demand to produce a given workability as illustrated in Figure 2.1 Increased difficulty in controlling entrained air content
Transport	Loss of water by evaporation Loss of workability
Placing, finishing and curing	Loss of water by evaporation Loss of workability Increased rate of setting Increased tendency to plastic shrinkage cracking Increased peak hydration temperature leading to increased tendency to cracking and lower long-term strength
Long term	Lower strength Decreased durability Variable appearance

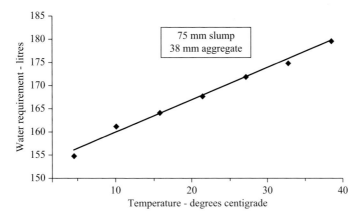

Figure 2.1 Effect of mixing temperature on water requirement for a given slump (after ACI 305) (courtesy of American Concrete Institute, 2014).

concrete, leading to potential problems with reinforcement corrosion. Onshore winds carry chloride-laden spray onto the facades of buildings and structures.

Reinforcement corrosion beneath bridge bearings

There were many instances of poor performance. Although this was tacitly acknowledged in the publication of The CIRIA Guide (Construction Industry Research and Information Association, 1984), very few examples found their way into the technical literature. An illustrative example is the piers of a highway bridge crossing a creek in a Middle Eastern country. The reinforced concrete piers were of the form shown in Figure 2.2, with castellations carrying the bearings beneath steel box girders.

The steel box girders were fabricated abroad and these, along with the bearings, were substantially delayed by congestion in the port. This meant that box-outs, of the form shown in Figure 2.3, had to be formed in the top of each pier to accommodate later installation of the holding-down bolts for the bearings.

The bridge was eventually completed, but within two or three years of its opening, cracks appeared in the sides of the piers just beneath the positions of the bearings, as illustrated in Figure 2.4. When some of the concrete in this region was carefully removed it was found that the congested reinforcement was heavily corroded. After an investigation of the extent and degree of the corrosion to the reinforcement, it was decided to demolish the castellations and rebuild them. This involved cutting slots for steel needle beams beneath the bearings and providing temporary support for the bridge deck using steel towers supported off the pile caps. The work was completed with only short temporary closures of the highway, mainly during the night.

Investigations showed elevated chloride concentrations were present in the concrete in the region beneath the bearings. It was concluded that brackish water had probably been used to cure the piers. Fresh water was in short supply and it would have been difficult to transport the necessary quantities to the required location in the middle of the creek. Curing water would have ponded in the box-outs at the tops of the castellations, in close proximity to the congested reinforcement beneath the bearings.

15

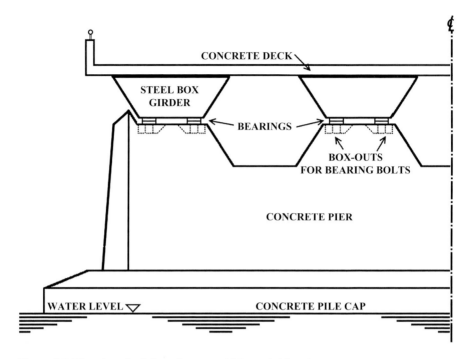

Figure 2.2 Elevation of reinforced concrete highway bridge pier.

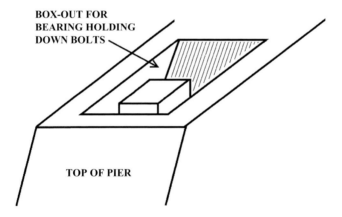

Figure 2.3 Box-out for bearing holding-down bolts in top of pier.

The fact that saline water had been used for curing was confirmed by sampling and testing of concrete adjacent to the construction joints at the base of the castellations. This showed elevated concentrations of chloride close to the top surface of the bottom lift (Kay *et al.*, 1982).

The provision of the box-outs was a direct consequence of the circumstances of the times – the bearings were delayed at the port. If the box-outs had not been

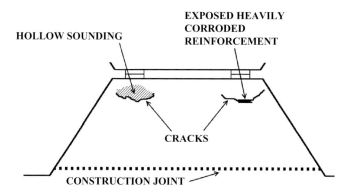

Figure 2.4 Cracks in side of pier.

constructed the effect of brackish water curing would have not been so severe and corrosion of the reinforcement, although still likely, would have been delayed for a more extended period.

Wider ramifications

The particular case described above was by no means an isolated incidence of early deterioration of a concrete structure in the region at that time. It soon became apparent to consultants and contractors with projects in the region that something was seriously wrong and efforts were made to determine the cause or causes. Initially, efforts were made to improve the knowledge of those working within the industry and to try to improve working practices and the quality of materials that were being employed. An example of this was the handbook on concrete practice produced by the consulting firm Halcrow and given to each member of its professional staff (Kay *et al.*, 1979).

Subsequently, a number of mainly British consultants and contractors worked together under the aegis of CIRIA (the Construction Industry Research and Information Association) to develop a common approach. This resulted in the publication of what became known throughout the region as 'The CIRIA Guide' (Construction Industry Research and Information Association, 1984). The guide was widely welcomed and accepted and was quoted in the documents of major specifying authorities in the area.

Another major influence in the quest to improve standards was the series of conferences on concrete deterioration organised by the Bahrain Society of Engineers starting in 1985 (Bahrain Society of Engineers, 1985). These conferences were initially held every three years, first in association with CIRIA and later the Concrete Society.

The CIRIA Guide was substantially rewritten and published by the Concrete Society and CIRIA (2002)[12]. Subsequently, a companion volume on the design of structures in the region was published by the Concrete Society (2008). The durability provisions of the Design Guide are a good indicator of the progress that has been made in developing specifications tailored specifically for the region. An extract from these is shown in Table 2.3. It is difficult to make direct comparisons between these recommendations and the requirements of CP 110 (British Standards Institution,

Table 2.3 Extract from the durability provisions in Table 1 of the 'Guide to the design of concrete structures in the Arabian Peninsula' (courtesy of Concrete Society, 2002).

Exposure	Nominal cover to reinforcement – mm								Cement[1]
	40+Δc	45+Δc	50+Δc	55+Δc	60+Δc	65+Δc	70+Δc	75+Δc	
Aggressive (wet, rarely dry)	N/A	N/A	C60 0.35/400		C50 0.4/380	C40 0.45/360			C
	N/A	C60 0.35/400		C50 0.4/380	C40 0.45/360				D
	C60 0.35/400		C50 0.4/380	C40 0.45/360					E
Severe (moderate humidity)	N/A	N/A	N/A	C60 0.35/400		C50 0.4/380	C40 0.45/360		C
	N/A	N/A	C60 0.35/400		C50 0.4/380	C40 0.45/360			D
	N/A	C60 0.35/400		C50 0.4/380	C40 0.45/360				E
Extreme (cyclic wet and dry)	N/A	N/A	N/A	N/A	N/A	C60 0.35/400	C50 0.4/380	C40 0.45/360	C
	N/A	N/A	N/A	N/A	C60 0.35/400	C50 0.4/380	C40 0.45/360		D
	N/A	N/A	N/A	C60 0.35/400	C50 0.4/380	C40 0.45/360			E

Note 1: Cement types are as follows:
Type C – PC/pfa (65-75%/35-25%) or PC/ggbs (40-50%/60-50%):
Type D – PC/pfa (50-64%/50-36%) or PC/ggbs (25-34%/75-66%);
Type E - Triple blend PC/fly ash/silica fume (55%-70%/35%-25%/10%-5%) or Triple blend PC/ggbs/silica fume (30%-45%/60%-50%/10%-5%)

1972) shown in Table 2.1, which were widely used for the Middle East in the mid to late 1970s. Looking at the 'Very severe' exposure from CP 110 gives a minimum cover of 60 mm for this situation, but no guidance is given for concrete proportions. This compares with the 'Extreme' exposure in the Gulf Design Guide (Concrete Society, 2002), where a minimum cover of 65 mm (nominal cover of 55 mm plus a tolerance, Δc, of 10 mm) requires the use of a triple blend mix with a minimum strength class of C60 (cube), a maximum water cement ratio of 0.35 and a minimum cement content of 400 kg/m^3. This is a concrete of much higher quality than those given in the CP 110 table.

Case Study B: Softwater leaching in a service reservoir

Introduction

It is well known that moisture containing dissolved salts, typically chlorides or sulfates, can have a large impact on the durability of concrete structures. The damaging effects of pure water, or even soft water, are less generally appreciated.

A reinforced concrete service reservoir was constructed in a West Midlands (UK) town in the period 1925–1928. Initially, the reservoir performed very well, but gradual deterioration was noted from 1954 onwards. Some remedial works, including minor patching and coating, were carried out in 1971–1972. By 1984, repair or replacement was being considered by its owners and an evaluation report was commissioned from consultants in 1986. The study considered various repair options but concluded by

18

recommending that the reservoir should eventually be demolished after the rehabilitation of a neighbouring reservoir.

These recommendations were not immediately implemented, as further structural and condition assessments were carried out in 1986. The details and findings from this study are described in the following.

Details of service reservoir

The reservoir is roughly trapezoidal in plan, as shown in Figure 2.5. It is located in a residential area on a sloping site bordered by roads to the southeast and southwest, a school campus to the northwest and housing to the northeast. The plan area is approximately 5,720 m², with inlet and outlet arrangements at the northeast and southwest ends, respectively. The interior length is 104.8 m and at the broader, southwest end the interior width is 70.43 m. It continues at this width for a distance of 20.7 m and from this point it tapers, with the side walls converging to give a width of 20.12 m, at the inlet end of the reservoir. The maximum water depth is 6.1 m, giving a storage capacity of 31,823,000 litres.

The construction is of a hybrid form, with a structural steel skeleton framework encased by reinforced concrete, the materials being designed to act compositely. This was a patented form of construction developed by Harry C. Ritchie (Gould, 2003). The external walls are made up of horizontally arched reinforced concrete panels that span between composite reinforced concrete and steelwork counterforts, which act as vertical beams transferring the outward water pressure into the roof and floor slabs. The reservoir floor and roof slabs are of reinforced concrete construction. The roof slab is carried by an arrangement of steel beams encased in reinforced concrete, which in turn is supported by a grid of similarly constructed columns at 4.19 m × 5.03 m centres.

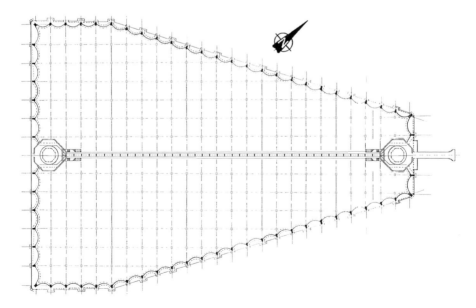

Figure 2.5 Plan of service reservoir.

Table 2.4 Water analysis carried out in 1994.	
Quantity	Value
pH	6.99
Conductivity	499 μS/cm
Alkalinity (as HCO₃)	115 mg/l
Residual chlorine	0.054 mg/l
Chloride	48.1 mg/l
Nitrite (as NO₂)	<0.01 mg/l
Nitrate (as NO₃)	19.9 mg/l

The columns are around 6.5 m long and have a cross-section of 254 mm × 254 mm. Each column contains a centrally located 5 in. × 4½ in. × 20 lbs/ft rolled steel joist and four main 20 mm diameter reinforcing bars. Links are 6 mm in diameter at 120 mm centres vertically with cover of 25 mm. The details of the concrete used in the columns are uncertain, but published papers on the reservoir suggest that it had an aggregate to cement ratio of 4½ to 1 with ½ in. down aggregate and 350–400 kg/m³ of rapid hardening Portland cement. The range in cement content reflects uncertainty about whether the concrete was batched by mass or by volume (the normal practice at that time). The water cement ratio was thought to be less than 0.5, with average 28 day cube strengths of around 40 N/mm².

Analysis of the water supplied to the reservoir in 1994 gave the values shown in Table 2.4. These values are average figures over a 10 month period.

Service history

The reservoir was first filled in May 1928 and gave good service over the first 26 years. The aggressive nature of the water was recognised as early as 1932 when badly corroded metal screens had to be removed. In 1953 it was first reported that the concrete had been affected. The surface could be removed by rubbing to expose the aggregate beneath. Over the next twenty years there was further gradual superficial deterioration, leading to sprayed concrete repairs to the inside wall surfaces in 1971 and sprayed concrete repairs to slab soffits and beams and general patching in 1972. A bituminous coating was also applied at this time.

In the following twenty years to 1995 there was steady deterioration, with breakdown of coatings and progressive softening of the surface. This resulted in the onset of corrosion in the reinforcement in the most vulnerable areas, particularly at roof level. The basic structural steel skeleton was still thought to be sound. It was at this stage that the study described here was commissioned.

Assessments

Both structural and condition assessments were commissioned, but this section will concentrate on the condition assessment of internal members in contact with the water.

A visual inspection was first carried out. This confirmed previous reports of softening of the concrete surface. This had occurred throughout the reservoir below water level, affecting columns, the central division wall, baffle walls and floor slab. It was also noted to a lesser extent on the sprayed concrete coating applied to the external arched walls. The aggregate was prominently exposed in the affected elements, with much of the cement matrix having been leached out, leaving only fine sandy material that could easily be scratched away to a depth of around 2–4 mm. Rusty nodules and rust staining were observed on the lower lengths of around 25% of the columns.

Representative areas of members with these features were subjected to more detailed examination to try to determine the condition of the materials behind the affected surface and to try to understand any underlying deterioration processes. The test methods included:

- Visual appraisal
- Surface hardness (Schmidt hammer)
- Location of reinforcement by covermeter
- Carbonation depth
- Half-cell potential
- Breaking out to check condition of reinforcement

A column in the South Compartment had some exposed coarse aggregate at its surface and rust stains close to its base. The mortar was soft and could be scraped away between the coarse aggregate. This condition applied at the base of the column and at mid column height. At the top of the column the concrete appeared to be hard and dense. Hardness was measured at the top and bottom of the column and near the mid point of its height, and gave the following values:

- Bottom of column   22
- Mid column   17
- Top of column just beneath top water level   41
- Top of column just above top water level   43

These values approximate to strengths in N/mm^2 and indicate that the column concrete had retained its original strength at the top of the column, and at the bottom the surface had lost about 50% of its strength due to leaching. A further hardness test was carried out at the base after removing approximately 10 mm depth of surface concrete. This gave a value of 27, suggesting that there had been progressive softening of the concrete from the outer surface.

Half-cell potential measurements using a silver/silver chloride half-cell were carried out at 0.5 mm intervals up the column. The electrical potential was −160 mV above top water level and decreased progressively to −490 mV near the base of the column. This pattern of values is indicative of an increasing risk of corrosion from top water level to base, although this is likely to be mitigated by lower oxygen availability at the lower level.

A carbonation depth determination using phenolphthalein gave a value of around 5 mm but did not display a distinct carbonation front – that is, an abrupt transition from colourless to pink stained concrete. This suggests a gradual loss of alkalinity in the transition zone due to soft water or acidic leaching.

Reinforcement was exposed locally adjacent to the area of rust staining. The reinforcement, which had a cover of approximately 20 mm, was found to be in good condition other than for an area of black corrosion product approximately 100 mm above floor level. The corrosion product gradually turned orange brown on exposure to the atmosphere.

Effects of soft water

It was concluded from the assessment of the column and other members in contact with water in the reservoir that long-term leaching of the cementitious matrix had led to depassivation of the reinforcement. Corrosion had taken place at a slow rate governed by the low availability of oxygen. Corrosion products had been able to leach back out from the corrosion site through the porous concrete, with ultimate deposition on the surface. This, taken with the fact that low oxygen corrosion does not produce the same volume of corrosion product as atmospheric exposure, meant that there was little expansive disruption of the cover concrete. That is, corrosion of the reinforcement had taken place, with little of the cracking and spalling that is normally associated with this process.

Because the loads on the structure were carried mainly by the internal structural steelwork, it was concluded that the reservoir was still structurally sound. Proposed remedial works included removal of areas of softened concrete, replacement of heavily corroded bars, and reinstatement with a polymer-modified cementitious repair system.

Case Study C: Thaumasite sulfate attack to bridge foundations

In February 1998 during bridge strengthening works, deterioration was discovered to the foundations of a 30-year-old motorway overbridge. The bridge deck is supported on piers at the central reservation and abutting the hard shoulders. Each pier has a set of three slender columns 450×750 mm in cross-section and 13 m long, resting on a spread footing approximately 5.5 m below ground level.

The concrete surface was sufficiently soft that corners of the column could be broken away by hand. Initial tests proved inconclusive as to the cause of the problem, chiefly because the soft reaction products were lost in the sampling process. Subsequent samples were subsequently subject to analysis (e.g. X-ray diffraction, XRD), which identified thaumasite sulfate attack (TSA).

TSA is distinct from conventional sulfate attack in that the calcium silicate hydrates react with sulfates to form calcium sulfate carbonate silicate hydrate instead of reacting with calcium hydroxide and calcium aluminate hydrates to form gypsum and ettringite. Prior to this discovery the known incidence of thaumasite sulfate attack (TSA) in the UK had been limited to a small number of cases in non-structural concrete exposed to cold wet conditions (Thaumasite Expert Group, 1999).

There are several risk factors in the occurrence of TSA:

- Exposure to high levels of sulfate
- Very wet, cold conditions
- Source of calcium silicate (e.g. Portland cement)
- Source of carbonate (e.g. limestone aggregate or bicarbonate in the groundwater)

In the case of this bridge, the columns were formed from *in situ* reinforced concrete cast in formwork within a cutting through undisturbed Lower Lias clay. Excavated Lower Lias clay was stockpiled and later used to backfill around the columns and form a 2 m embankment either side of carriageway. Oxidation of the pyrite in this reworked clay is thought to have led to acid sulfate conditions being developed; first in the backfill and subsequently in the groundwater percolating the backfill, and collecting in the excavation next to the pier.

The presence of TSA in the structure was significant because it meant that concrete designed to the contemporaneous guidance at the time of construction would not have adequate sulfate resistance. Furthermore, site investigation prior to construction would not have identified aggressive conditions or considered the potential risk of acid sulfate conditions being generated during construction by excavation and backfilling.

The guidance was subsequently changed to address these shortcomings (Building Research Establishment, 2005). Under the combinations of most aggressive conditions and long design life, the concrete would have to meet limits on the mix proportions, including:

- Minimum percentages of fly ash or ground granulated blastfurnace slag
- Maximum water/cement ratio and minimum cement content
- Additional protective measures, e.g. sacrificial layers or coatings

The structure was subject to extensive investigation, followed by remedial work as outlined in Table 2.5.

A condition survey was carried out in which softened areas were mapped and the depth of expansion and softening were measured for each face of the columns. Concrete cores and dust samples and fragments were subject to petrographic examination, scanning electron microscope (SEM) microprobe analysis, compressive strength testing, and chloride and sulfate determination (Floyd and Wimpenny, 2003).

Table 2.5 Summary of investigation and remedial work.

Action	Purpose
Stage 1 – confirmation of deterioration extent and cause	
Sampling and logging excavation	Confirm soil type and sulfate conditions
Condition survey (mapping area and depth of softening)	Determine extent of deterioration
Cores for petrographic examination, SEM work and chemical analysis	Confirm of TSA from reaction products Determine sulfate and chloride profiles
Stage 2- remedial work	
Propping of structure	Ensure safety from dead and live loading, e.g. heavy goods vehicle impact
Cutting and removal of existing piers	Remove deteriorated concrete and provide samples for future testing and examination
Recasting new piers	Replace vulnerable concrete with a concrete mix and additional protective measures meeting the new guidance

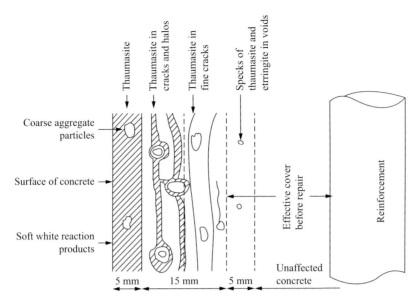

Figure 2.6 Schematic representation of TSA-affected concrete.

The area of softening was determined by hammer tapping and the depth was assessed by non-percussive drilling. There was generally no attack within 1 m of ground level, partial attack in the form of patches or bands of softening up to 500 mm in size approximately 2 m below ground level, and attack across the full area of all the faces below 4 m depth.

The attacked concrete was observed to have a soft, white pasty layer, typically 5 mm in thickness, but no visible cracking (Figure 2.6). Below this layer, the concrete was found to have cracks running subparallel to the surface, and the coarse aggregate particles were surrounded by white 'halos' of reaction products. The frequency and size of the halos reduced with depth below the exposed surface such that, typically at 20 mm depth, only occasional white specks of reaction products were present in pores, without any physical disruption to the cement matrix.

The reaction products, which were initially very soft, became hard and friable after being exposed for several days. Rust staining was present within some of the attacked areas, and breakouts revealed deep pitting corrosion to the reinforcement, representing 30% loss of section, and black and green reaction products that turned brown on exposure characteristic of anaerobic chloride-induced corrosion (Figure 2.7).

Sulfate and chloride determinations from 10 mm depth increments of a 100 mm diameter core are shown as profiles in Figure 2.8. The depth of TSA given by the depth of cracks infilled with thaumasite from the petrographic examination is also indicated. It can be observed that the depth of TSA approximately corresponds to the point at which sulfate values exceed approximately 5% SO_3 by mass of cement. The chloride profile shows a peak of over 0.8% chloride ion by mass of cement, which is approximately 5–10 mm deeper than the depth of TSA. This chloride content exceeds the corrosion threshold (Comité Européen du Béton, 1992) and, together with the

Figure 2.7 Bridge pier with white TSA reaction products, rust staining and pitting corrosion to reinforcement.

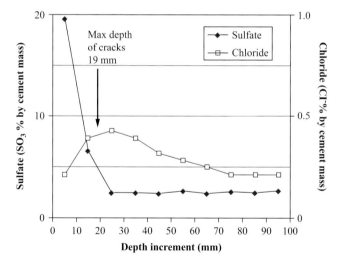

Figure 2.8 Sulfate and chloride profiles.

evidence of corrosion to the reinforcement, indicates that chloride-induced corrosion may be an additional problem in areas where TSA is present.

The bridge deck was propped as a precaution. The maximum net loss of cross-section of the columns due to softening was found to be 33 mm. Although there was no immediate risk of buckling failure to the columns, in view of the possibility of further deterioration during the remaining service life it was decided to replace the affected lower portion of the columns and recast a new spread footing above the old base.

The problems experienced at this site resulted in the following guidance 'It is particularly important in aggressive ground conditions to avoid the situation where a backfilled excavation acts as a sump, ponding water against the structure. This would be particularly aggressive to concrete if the backfill contains sulfate or sulfate bearing material (e.g. pyritic clay)' (Building Research Establishment, 2005).

Case Study D: Acid attack to a bund at a water treatment works

The bund was constructed in approximately 1991 for the containment of acid spillages at a water treatment works. The bund comprises reinforced concrete walls 400 mm and 800 mm in thickness and 1.15 m in height, enclosing an area of floor 8.5 m long and 7 m wide (Figure 2.9). The floor is formed from a reinforced concrete slab 600 mm in thickness and includes a channel alongside one wall leading to a sump to provide for minor spillages and washing down. There are two sulfuric acid tanks within the bunded area, supported on 1 m high reinforced concrete plinths.

The area was subject to a spillage of 98% sulfuric acid in 1999. The acid was contained within the bund and was neutralised before being removed after approximately 9 months. Damage was noted to the floor and walls and a condition survey was carried out to determine the extent of damage and establish potential remedial measures. The survey involved visual inspection, hammer tapping, non-percussive drilling to assess the depth of any softening and remove concrete dust at 5–30 mm, 30–55 mm and 55–80 mm depth increments. These samples were subject to chloride and sulfate testing in accordance with BS 1881, Part 124.

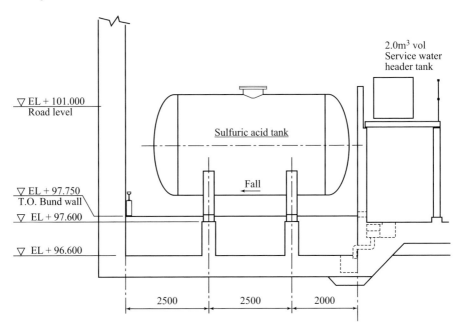

Figure 2.9 Elevation showing bund and sulfuric acid tank.

A fibre-reinforced liner was present on three of the bund walls, the tank plinths and the floor. The liner to the floor was severely deteriorated, with 80% of the concrete surface exposed, and the residual area of liner had de-bonded from the concrete. The liner to the bottom 60 mm of the walls had completely deteriorated to expose the concrete. Above this area the liner showed white staining indicating acid damage due to splashing and the residual liner had de-bonded to a height of approximately 200 mm from floor level.

Where the liner was present the concrete was reasonably sound, with a depth of softening (assessed by drilling) of 2–4 mm, but elsewhere the concrete had deterioration in the form of exposed coarse aggregate due to loss of the cement paste. The exposed aggregate appeared to be limestone coarse aggregate, with some flint particles in the coarser fraction of the fine aggregate. The depth of softening by drilling was 4–9 mm. There was no detectable hollowness or cracking.

The channel and sump were more heavily deteriorated than the rest of the bunded area (Figure 2.10). Reinforcement was exposed, indicating at least 30 mm depth of concrete had been lost based on measured cover values. In general, the reinforcement had minimal loss of section, as evidenced by the presence of ribs on the deformed bars, although some bars were more heavily corroded within the sump (Figure 2.10).

The chloride values from the dust sample were less than 0.1% by mass of binder (assuming a 14% binder content), indicating a negligible risk of chloride-induced reinforcement corrosion (Quillin, 2001). High sulfate values of up to 25% SO_3 by

Figure 2.10 Acid attack to sump showing exposed reinforcement.

mass of binder were present in the 5–30 mm depth increment, consistent with prolonged exposure of the surface to concentrated sulfuric acid (Figure 2.11). There was
some limited evidence that the sulfate content in this depth increment reduces with
increasing height above the floor level, with the value reducing from 22% at 50 mm
above floor level to 9% at 85 mm above floor level. The normal limit on built-in sulfate
content in concrete is 4% SO_3 by mass of binder, and values above 5% SO_3 indicate a
risk of sulfate attack. The latter would apply to concrete up to approximately 45 mm
deep within the walls.

The nature of the grey fibre-reinforced liner is not known. However, it could be
reasonably expected that the floor would be more vulnerable to damage than the walls
due to the combination of damage to the liner due to trafficking in service and the
greater duration of contact and head of spillage lying on the surface. This is consistent
with the reduced damage higher up the walls and the increased damage in the channel and sump. Once the 98% sulfuric acid penetrates through the liner to the concrete
it will attack the exposed concrete surface, removing calcium and leading to breakdown of the calcium silicate hydrates and calcium aluminate hydrates that provide
strength to the cement paste. Gradual softening at the interface with the liner will lead
to it de-bonding, allowing further attack to the concrete surface. Limestone aggregate
exposed by loss of the cement paste may temporarily buffer the reaction where the
residual de-bonded liner prevents the acid from being readily replenished. However,
reduced support from the substrate would be expected to result in cracking to the liner
and ultimately to complete loss of integrity.

Acid conditions are commonly expressed by pH. However, this is uninformative
about the nature of the acid or its concentration. For example, sulfuric acid at 4%
concentration would have a pH of virtually zero and be indistinguishable from acid
at 98% concentration (Eglinton, 1987). No Portland cement-based concrete will resist

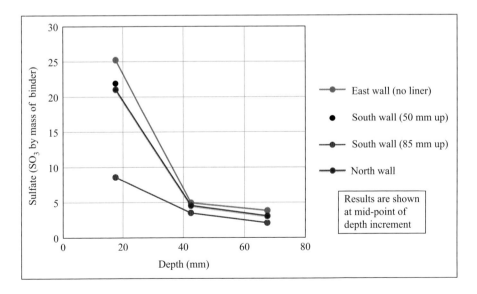

Figure 2.11 Sulfate profiles within walls.

pH values below about 6 (Eglinton, 1987) and a liner is recommended for water storage and transportation structures where the pH of the water is less than 5 (Building Research Establishment, 2005). The precise conditions are important in selection of the protection. The liner in this case proved unsuitable for the chemical exposure and possibly the physical conditions, such as trafficking. The maintenance and age of the liner and the housekeeping procedures for dealing with spillage could also have been factors. Maintaining infrastructure and procedures is a particular challenge where these only come to the fore in very rare times of emergency. In this case, despite the damage to the liner and concrete, the bund served its function in containing the spillage.

Liners suitable for the exposure conditions include trowel-applied liners with an epoxy resin base, reinforced with a woven glass cloth. This would be expected to have a life of at least 10 years.

Case Study E: Fire – the Buncefield incident

A massive explosion and fire at the Buncefield oil terminal in Hertfordshire, UK, on 11 December 2005 caused significant damage to the terminal and neighbouring buildings. Significant environmental damage was also caused by leakage of petroleum products and contaminated firewater through failed secondary containment bunds and tertiary containment measures. Because the groundwater was contaminated over an area in excess of 1 ha, the incident was defined as a major accident to the environment (Tarada and Robery, 2014a).

Concrete is well known for its excellent fire-resisting qualities, so it was surprising therefore that the concrete secondary containment was reported to have failed, following the series of explosions that damaged 22 primary containment fuel storage tanks. As well as the structural damage arising from the blast wave from the initial large explosion, which destroyed most of the site and surrounding buildings, the resulting fire then raged for five days, as the fire service battled to control the fires and prevent unaffected areas also becoming involved.

The burning fuel that spilled out of the primary containment and into the bunds from the ruptured tanks formed 'pool fires', which were eventually controlled by a combination of foam and water. The bigger problem was the fuel that remained inside the damaged tanks, which proved more difficult to control and the tanks were often located centrally within the bunds and difficult to reach by the fire service. A combination of water and foam was used to control the fires, estimated to be 53,000,000 litres of water and 786,000 litres of foam and assorted combustion/decomposition contaminants, collectively known as 'firewater'.

As the bunds with burning tanks filled with water and foam, the fire service sought to pump the excess firewater out of the bunds where the fire still raged, as these were filling and would over-top the bund walls. Part of the emergency plan was to pump the firewater into bunds where the fire had been extinguished and the largest and most recently built bund on the site was chosen for this use, to the north of the site.

Unfortunately, the bund did not fulfil its intended function of containing the firewater and residual fuel until they could be safely removed, which is normally planned to be an 8 day retention period post-incident (Mason *et al.*, 1997). Large quantities of fuel and firewater escaped from this bund and from other, older bunds. Some bund failures were inevitable, as the severity of the fire in places had caused the concrete to decompose and steel to melt. At other locations, where the fire was not as intense, both good and bad practices were identified from the later forensic examinations. It is the performance of these bunds that is examined here.

By definition, a bund is commonly described as a construction with walls and a base that is built around an area where potentially polluting materials are handled, processed or stored in a primary containment system (e.g. a storage tank), for the purpose of containing any unintended escape of material from that area until such time as remedial action can be taken (Mason *et al.*, 1997).

Traditionally, bunds have been built from a variety of materials including concrete-filled reinforced blockwork, compacted earth walls and bases, and reinforced concrete. For Control of Major Accident Hazards (COMAH) sites containing petroleum products, the choice of construction material is much more limited to ensure satisfactory performance in a fire. The material choice narrows to earth bund or concrete bund construction or, as is commonly found with older facilities, earth bases with concrete walls. The requirement is that all materials achieve an impermeability to the contained liquids equivalent to a material of 1 m thickness with a permeability coefficient of 1×10^{-9} m/s, which is deemed by industry to represent best containment practice for clay liners (Report C164) (Mason *et al.*, 1997). These requirements apply not only to the materials of the walls and bases, but also to the ancillary materials used to build the bunds, including joints and pipe penetrations.

Because of the requirement to contain any escaping liquids or firewater for a minimum period of 8 days, to allow sufficient planning time following an incident to arrange extraction by tanker or other measures, the bunds should be designed and built as liquid-retaining structures. CIRIA C164, published in 1997, references the now superseded BS 8007 (British Standards Institution, 1987) and recommends water-retaining design and construction practices (Mason *et al.*, 1997).

The investigations at Buncefield identified that bund performance depended on the time of design and construction, with the general trend that the older bunds performed better than the most recent one (Figure 2.12).

- The first bund at the site (the 'Main Bund') was built in the 1960s and comprised reinforced concrete walls and a sunken earth floor (i.e. the bund floor was below ground level), with dowels across the joints and standard water-stop technology of the time (commonly found in water-retaining structures), namely a copper sheet with a profiled section that was embedded in the concrete either side of the vertical construction joint in the wall. *These performed very well and apart from local surface spalling, the joints were effective at containing liquids.*

- In the 1970s more tanks and bunds were added to the site (Bunds D and E), again incorporating a sunken earth floor, but using central polyvinyl chloride (PVC) 'dumbbell' waterstops across vertical dowelled joints in the wall. As part of a risk assessment, these had been upgraded by bolting steel plates onto the wall to offer protection against leakage and heat from a fire. *These bunds also performed well and were effective at containing liquids, and the PVC waterstop was undamaged.*

30

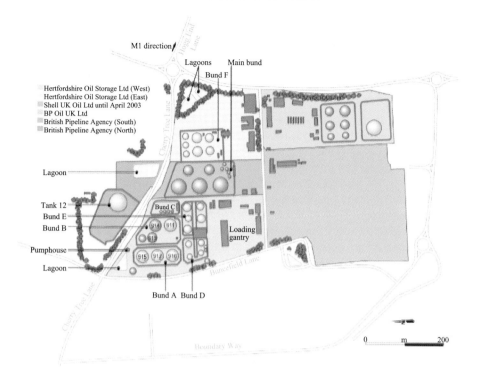

Figure 2.12 Buncefield site layout (based on Tarada and Robery, 2014b).

- In the late 1980s, Bunds A, B and C were added, Bund A being the seat of the first explosion, and again they had sunken earth floors, but deviated from earlier construction as waterstops were not provided at the dowelled vertical joints, with mastic sealant gunned into the surface of the joints. *Because the mastic was not fire-resistant, it was consumed by the fire, allowing fuels and firewater to escape through the joints.*
- Finally, the Tank 12 bund was completed in 2002, comprising a concrete bund floor and wall system, with the floor being level with the ground for around 50% of the bund. The concrete floor was built using a traditional ground-bearing design, reinforced with fabric and, between dowelled construction joints, it was sawn into bays of nominally 6 m square (variable). The walls were built as freestanding sections, with no dowels provided between adjacent sections of wall. No waterstops were used in the floor or walls, with only mastic sealant applied into the joints. Furthermore, because the bund was not rectangular, it had joints in the walls at shallow angles of between 15° and 45°, with the joint positioned in the centre of the angle. *Because the mastic was not fire-resistant, it was consumed by the fire, allowing fuels and firewater to escape through the joints in the walls and floor slabs. Also, as the walls expanded from heat arising from the fires, the dowel-free sloping wall joint resulted in the sections pushing outwards, so compounding the issue.*

31

It has to be stressed that the duration and severity of the incident meant that nothing would have prevented some pollution of the ground, in situations where 'jet fires' from leaking fuel escaping from fractured tanks and pipes exposed the concrete to very severe heat. As noted above, some concrete decomposed and effectively melted, along with the steel reinforcement. In other locations, the depth of severe heating and decomposition (above 600°C) was found to depths approaching 150 mm in the floor slab, resulting in the floor lifting off the sub-base beneath.

However, the findings about the lack of any provision of waterstops and dowels, particularly in the newest construction, built under a design and build contract, came as a significant surprise, given the robustness and detail of the CIRIA Report C164 recommendations. It seemed the walls had been designed without due consideration of the effects from a fire in the bund.

The walls to the Tank 12 bund were also found to have been built with through-wall, removable formwork ties at low level, just above the kicker to the wall base. The tie-bolt holes had not been cleaned out and filled with grout in the normal way for a water-retaining structure. Instead, the surface cone (formed by the removable plastic spacer used) was plugged at the surface, so that when the wall was exposed to heat from the fire, thermal expansion caused the plugs to drop out, allowing firewater to escape.

The construction generally did not follow another key recommendation in CIRIA Report C164, namely to avoid pipe penetrations through the wall, or if they are provided, to ensure they are watertight using standard technology used in swimming pools, such as puddle flanges. Instead, many bund walls had over-sized holes that were provided with flexible gaskets to make a watertight seal. Unfortunately, the evidence was the gaskets were not resistant to heat from the burning fuels.

Following completion of the forensic engineering investigations, the Environment Agency commissioned CIRIA to review and update its 1997 Report C164. A new version was published in 2014, incorporating the findings from the investigations, updated safety and environmental standards for fuel storage sites, and guidance from the Buncefield Standards Task Group (Buncefield Standards Task Group, 2007) (BSTG). Report C736 provides detailed guidance on evaluating existing bunds for adequacy and fitness for purpose (Walton, 2014).

In summary, the key technical points include the following:

- Design bund walls and floors as a liquid-retaining structure to BS EN 1992-3 (British Standards Institution, 2006) or equivalent, using best practice from the water industry (UK Water Industry Research, 2011).
- Ensure all bund walls and floors have properly designed and carefully built joints, with central waterstops that should be stainless-steel sheet at sites in the highest risk category.
- Avoid removable through-wall tie-bolt methods for fixing formwork and use cast-in types or suitable framing that does not use through-wall tie bolts.
- Do not route pipework through the concrete floor slabs, but if it is not possible to avoid this, provide proper puddle flange connections to form a watertight seal.
- Set up a regular inspection maintenance regime as befits any safety-critical element of the facility.

- For existing facilities, carry out a baseline survey to find out how bunds have been built and whether there has been a change in use/risk. If deficient, address any shortfall by extending or upgrading the secondary and/or tertiary containment.

Case Study F: Abrasion in aggregate storage bins

The aggregate storage bins studied here were constructed in the 1960s to store different sizes of crushed rock at a granite quarry. The facility is 30 m long and 8 m in height and predominantly formed from reinforced concrete. The perimeter walls are 200 mm thick and there are similar internal walls to form 12 bins of various capacities (Figure 2.13). The bins are supported above ground level on 150 mm square columns formed from concrete-encased steel stanchions. The feed from the crushing plant is by an overhead conveyor and the discharge from the conveyor is intended to be directed away from the sides of the bins by hardened steel deflector plates, which can be replaced periodically.

A condition survey in 1997 found some deflector plates had abraded through their full thickness, allowing the aggregate flow to impact on the concrete. Severe damage was noted to six of the bins, including erosion of walls, exposing the reinforcement and with the formation of holes up to 2 m in height. There had been unsuccessful attempts to cover these holes with steel plate. In places the structural cross beams and stiffening ribs were also completely eroded away (Figure 2.14).

The external faces of the facility, with the exception of the east, had numerous small areas of hollowness, spalling, exposed reinforcement and previous repair. This was attributed to carbonation-induced reinforcement corrosion. On the west elevation, holes abraded through the perimeter wall permitted oversize aggregate to spill out onto the roof of the tertiary crushing facility below.

As a result of the damage, the affected bins were no longer fit for service due to the risk of cross-contamination of aggregate sizes. There was also concern about structural stability and the future deterioration of the internal faces of the remaining bins and the external faces of the facility.

Three options were considered: reinstatement, making safe and demolition. The client chose reinstatement for continuing use. These works were significantly hampered by the limited access and the confined working space within the bins.

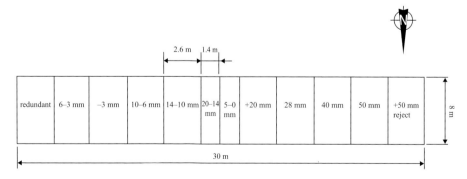

Figure 2.13 Schematic plan of the silo showing the aggregate sizes stored and selected internal dimensions).

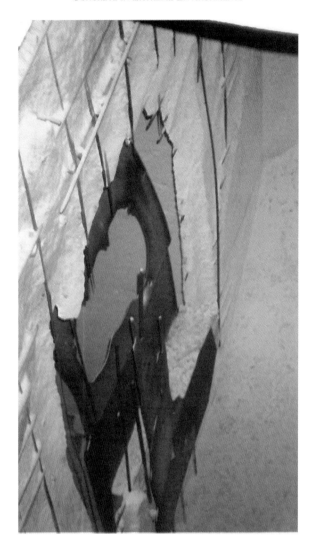

Figure 2.14 Typical abrasion damage to internal wall, showing exposed reinforcement, hole and previous repairs using steel plate.

There is limited codified guidance on the design and testing of concrete for abrasion and impact resistance, and this relates to more common exposure cases like floors and hydraulic structures (British Standards Institution, 2003; American Concrete Institute, 2007a,b). The guidance draws heavily on previous experience and typically makes recommendations about the following:

- Concrete strength
- Cement and water content
- Aggregate size and properties (e.g. resistance to fragmentation)

34

- Inclusion of specific materials (e.g. fibres and dry shake hardeners)
- Finishing and curing

The abrasion damage was remediated by applying sprayed concrete to the internal surfaces using the dry mix process. The dry components of the sprayed concrete were batched and mixed in a factory and incorporated 40 kg/m³ of steel fibres and silica fume for abrasion resistance (Table 2.6).

Lost reinforcement was replaced by new bars. Cross beams and other structural members were conventionally cast using high-strength concrete containing silica fume (Table 2.7). The concrete mix used the granite aggregate being processed by the plant. This aggregate has a 10% fines value (a measure of aggregate strength under crushing (British Standards Institution, 1990)) in excess of the 100 kN minimum value recommended for exposure to abrasion, such as in coastal structures (Allen, 1998). New deflector plates were to be installed after the refurbishment to reduce the impact on the silo walls from the discharging aggregate.

The corrosion damage to external surfaces was repaired by breaking out and reinstating using a proprietary repair mortar. A surface-applied corrosion inhibitor was applied to the surfaces to protect against continuing carbonation-induced corrosion.

Approximately one year after refurbishment, abrasion to the spray concrete repairs and a cross beam were noted in the oversize (+50 mm) bin. Within two years of the refurbishment, areas of the wall subject to direct impact from the aggregate

Table 2.6 Summary of specification for sprayed concrete.

Parameter	Value
Characteristic 28-day compressive strength (MPa)	50
Minimum cementitious content (kg/m³)	400
Maximum free water cement ratio	0.45
Proportion of silica fume (%)	5–10
Nominal maximum size of aggregate (mm)	8
Steel fibres:	
Minimum tensile strength (MPa)	800
Length (mm)	25–40
Minimum content (kg/m³)	40
Minimum bond strength (MPa)	0.5

Table 2.7 Summary of specification for conventionally cast concrete.

Parameter	Value
Characteristic 28-day compressive strength (MPa)	60
Minimum cementitious content (kg/m³)	370
Maximum free water cement ratio	0.45
Proportion of silica fume (%)	10
Nominal maximum size of aggregate (mm)	20

had abraded to over 30 mm in depth to expose areas of reinforcement. Unfortunately, recommendations to reconfigure the chute into the bin and install deflector plates had not been implemented.

In view of the nature of the aggregate being discharged into the bins, implementing an effective scheme to deflect the aggregate from direct impact on the concrete walls was a fundamental requirement. This case illustrates that, under extreme abrasion, it is necessary to consider additional measures instead of relying solely on the concrete. It is also important to have in place an inspection and maintenance regime to identify and address emerging problems, particularly if major repairs could be difficult to undertake due to access constraints.

Case Study G: Algae – experiences from Blackpool

As part of a PhD research project, Hughes (2014) studied part of the Blackpool promenade sea wall, which comprises a revetment forming part of the coastal defence scheme (Figure 2.15). Precast concrete armour units, appearing like giant steps, protect the promenade and road structures against erosion, with some parts being in service since 2006.

The low installation costs, low maintenance, high durability and high-quality finish of precast concrete units make this solution a popular choice for the construction of sea defences. The proliferation of concrete coastal protection has transformed sections of naturally dynamic coastlines into artificially static, hard substrata. The durability of marine concrete is arguably its principal property, and it should be capable of

Figure 2.15 Typical revetment units at Blackpool (www.slp-precast.com/submat/images/precast/blackpool1.jpg).

withstanding harsh conditions throughout the expected 100-year life of the structure to protect the coastline.

At Blackpool, the armour units were conceived as mass concrete structures that would not require conventional reinforcement. Principally, for handling and to reduce the concrete tensile strain, the concrete units incorporate macro-synthetic fibre reinforcement (Concrete Society, 2007). The installation in Blackpool appears to be one of the first uses of macro-synthetic fibre for concrete in a marine environment (Rieder, 2007). Thus, there was little information about how the physical properties of the synthetic fibres change with time of immersion or whether the long-term mechanical performance and durability may be affected.

Each precast concrete revetment unit measures 5 m × 35 m and contains 8 m³ of concrete. The 'as struck' concrete surface exposed to the sea was cast face-down against a steel mould. For reinforced concrete, the relevant exposure classes would be XS3, XF4, XC3/XC4 and XA to BS 8500-1 (British Standards Institution, 2006). Without steel reinforcement, the durability design simplifies to XF4 and XA, and even then BS 8500-1 permits declassifying the concrete as XF4 because it is 'frequently in contact with the sea'. The design parameter therefore becomes resistance to sulfate ions in seawater, typically taken to be 3,000 mg/l, with limiting concrete proportions given in Table A.12 of BS 8500-1.

The concrete mix proposed by the contractor comfortably passed the limits in Table A.12 and had a minimum cement content of 340 kg/m³, based on CEM IIIA cement (50% ground granulated blastfurnace slag), a compressive strength class of C35/45 and water/cement ratio of 0.45 (Hughes et al., 2013a). As the concrete did not contain reinforcement, marine-dredged fine aggregate was used in the concrete, without the concern about initial chloride ion content (chloride class) of the concrete that would exist for reinforced concrete, although this satisfied Class 0.2 of BS 8500-1. It was noted that the fine aggregate was only washed with seawater, after dredging to reduce the silt content and screening to remove debris, rather than being washed with fresh water.

Synthetic fibre types used in the revetment concrete include polyethylene macro-fibres, 40 mm in length, 1.4 mm wide, 0.105 mm deep, rectangular in section, and dosed into the cement at 3.5 kg/m³. In addition, polypropylene micro-fibres (monofilaments) of circular section and 22 μm in diameter were used at a recommended dosage rate of 0.91 kg/m³.

The installed units face westward and are subject to cyclic wetting and drying conditions in the tidal range and splash zone. At times, the concrete is exposed to harsh wave action, which is exacerbated by abrasion from fine aggregate and occasionally shingle and debris from the beach.

As Hughes et al. (2013b) identify, within a relatively short space of time, the moulded concrete face that had been cast against steel formwork, which is normally dense and smooth, became colonised by microorganisms, in particular slimy green algae. As a result, the flat surfaces of the revetments became very slippery when wet. As the flat areas made attractive walking and sunbathing areas for the public, the algae presented an immediate health and safety concern, prompting the City Council to try various measures to remove or retard growth of the algae. These measures included hypochlorite bleach to chemically kill the algae and high-pressure water blasting

(1,200 psi), dispensed from motorised units with multi-oscillating jets to remove the microorganisms. Both measures only provided a temporary respite, and with successive tides, the algae recolonised the revetments.

Hughes identified the algae as ulva and found that, as well as the obvious 'green uniform microbial lawn' of algae on the surface, filaments of the algae were found in the concrete to a depth of 20 mm, penetrating between aggregate particles and the cement paste, as well as within the macro-synthetic fibres themselves and in the mastic sealant (Hughes *et al.*, 2013c) used between the units. The question that remained is why the colonisation was so fast on an apparently smooth concrete surface. The answer was traced back to bacteria present in the concrete.

Most research in the field of bacterial activity has concentrated on the beneficial use of bacteria to create self-healing concrete (De Muynck *et al.*, 2010; M4L Programme, n.d.), where particular species of dormant bacteria in dense concrete can be activated by cracking and moisture precipitation to deposit carbonate into cracks and so seal any leakage. Surprisingly to concrete technologists, but less so to microbial specialists, despite the highly alkaline conditions in concrete, the bacteria revert to a dormant state and survive happily in the matrix.

The findings found that bacteria had been inadvertently introduced into the concrete from the marine-dredged fine aggregate. The beach sand used as fine aggregate, taken from the vicinity of the dredging site, was shown to be colonised by filamentous microbial growth (the existence of microbial populations that live on or within sand is well documented) (Hughes, 2013). The presence of the microbial population in concrete encourages colonisation of the concrete surface when it is exposed to seawater, such as in the tidal range and splash zone. Also, because the microbes are on a microsized scale, there is currently no provision in European standards for aggregates for limiting their presence, as they would not be detected from measurement of the organic content of the aggregate as is required by current standards (British Standards Institution, 2008).

Research using light and inverted microscopy, scanning electron microscopy (SEM) and energy dispersive X-ray analysis (EDX) of site specimens enabled the interaction to be understood further (Hughes *et al.*, 2013d). Algal fronds were observed growing between fine aggregate and/or fibre and the cement paste. Colonisation was observed entangled with and adhered to macro-synthetic fibres. This mechanism occurs when the bond is weakened as a direct result of physical activity of the organism, such as growth.

Loss of surface material from the concrete, either through biodeterioration or through mechanical means such as power washing, was found not to be the primary cause of degradation, but may exacerbate the condition. Visual inspections at the site over time showed that the synthetic fibres began to break away from the surface and then protrude from the concrete, anchored at one end. With the sway of the fibre back and forth when under water, this helped to weaken the bond of the fibre until it eventually broke away.

The penetration of algae into the concrete surface, particularly under and through fibres, accelerates degradation of the polymer and also offers algae refuge from hydraulic forces, whether these are tidal impacts or power washing. The presence of algae at the fibre/cement interface disrupts or distorts the fibre by its very growth. The organisms do not use the materials as nutrients, but simply weaken the mechanical bond between the aggregate or fibre and the cement paste matrix.

Algae can act as the focus for other biofouling organisms such as fungi and bacteria, so the deterioration process may gain momentum after the structure's condition has become suitable for the survival of one or more of the organisms. The algal filaments absorb and store seawater and associated nutrients during periods of submersion and from moisture found within the concrete. If the conditions (nutrients, UV, temperature and seawater) are favourable, the filaments will grow and extend over larger and deeper areas of the concrete. The effects of shrinking and swelling of the hydrophilic filaments during dry and wet conditions accelerate the mechanical biodeterioration of the bond between the aggregate or fibre and the cement paste.

The conclusion from the study has significance for some applications of marine-sourced fine aggregate. Although marine-sourced coarse aggregate can be washed effectively, fine aggregates are much more difficult to clean effectively and, with a large surface area, biofilms are likely to remain in place. It is likely therefore that where marine-sourced fine aggregates are used in concrete, a residue of microbial film may exist. Where any resulting concrete is to be used in the tidal range, with similar bacterial and algal activity as seen in Blackpool, the concrete is likely to be colonised rapidly by algae such as ulva. The growth of the filaments into the surface, around and between the aggregate and synthetic fibres, may also lead to weakening and erosion of the surface at a much faster rate than would be expected, even if the concrete surface is not hydroblasted to clean it.

The salient warning from this study challenges the common perception that deterioration of concrete structures usually starts at the surface and progresses into the cover zone; it is, therefore, the skin of the precast concrete units that is the major factor in the longevity of the concrete. Hughes' research shows the presence of algal filamentous growth already exists after casting the concrete, being present on the surface of fine aggregates, even if washed in seawater. This leads to microbial degradation of concrete in the tidal range, as studied at Blackpool, exacerbated when synthetic fibres are used in the concrete.

Case Study H: Mould growth on long-span bridges – a visual problem

Introduction

This case history concerns a study that was carried out when the visual appearance of large structures became a problem in the first few years after completion. 'Failure', in terms of visual appearance, is a relatively rare occurrence. Two long-span bridges were constructed in a Far East state to provide road and rail access from a commercial and residential centre to a new airport. One is a suspension bridge and the other cable-stayed. The cable support pylons of the suspension bridge are approximately 200 m in height, and those for the cable-stayed structure are approximately 150 m in height. There are few other tall structures in the locality and the pylons are a prominent feature of the landscape, being visible from a great distance. The concrete towers were constructed using a cement that contained a high proportion of blastfurnace slag to meet the durability requirements of the specification. This led to them having quite a light colour. They were part of a prestigious development and in some ways represented the gateway to the state and hence their appearance was important.

A few years after construction, some dark discolouration appeared on the surface of the towers. Some staining was apparent shortly after construction but became a lot worse within two or three years. The disfigurement appeared to be the result of some form of mould growth. Microscopic examination of 15 samples of concrete from the structures had revealed fungal growth on all samples. It had also been possible to detect fungal growth on most of the samples when they were cultured. Subsequent to the investigation of these samples a study was commissioned to investigate methods of cleaning and treating the surfaces to improve their appearance and to try to delay the subsequent regrowth of mould.

Discolouration

The discolouration first became noticeable on the cross girders of the pylons of one of the bridges within a year or so of completion. It appeared as large patches of black staining. On lower sections, where it was less noticeable to the general public (being below carriageway level), it occurred in horizontal bands. These horizontal bands tended to coincide with the joints between individual concrete placements where the porosity may have been greater and where there were small indentations on the surface, as shown in Figure 2.16. The appearance of discolouration was highly variable. For example, on one of the towers there were some larger areas at the base and near

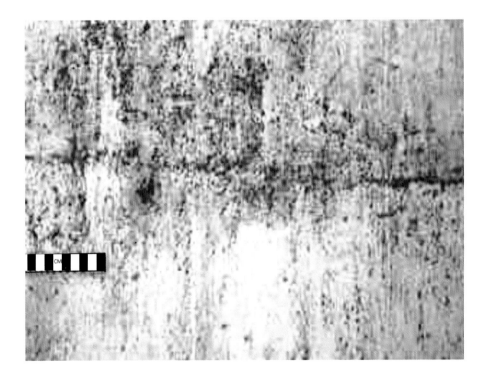

Figure 2.16 Mould growth at lift joint.

the top of the west face, with the remainder of the face being relatively free of staining. In this case, the areas of worst staining tended to coincide with areas of the pier that were in shade for at least part of each day.

Colonisation

There are several different types of organism that can cause disfigurement of concrete surfaces. These include bacteria, algae, fungi, lichens, liverworts and mosses. Mould is a type of fungus. It is a simple microorganism of single- or multi-cellular type in the structural form of a filament. In order to thrive, moulds colonise substrates where sufficient nutrients are available to them. They do not feed on concrete itself, but on dirt and dust, which often result from the presence of other organisms (being products of metabolism or decay).

One of the most important factors in determining the extent of colonisation is the intrinsic nature of the concrete surface, particularly its pH and ability to hold or shed moisture. In this respect, surface texture is important. Small crevices can trap moisture, dirt and other particulates and are not washed out by rainwater. They also tend to hold moisture longer as the surface dries out.

Environmental conditions are also important. The availability of sunlight is a factor, as most colonising organisms rely on photosynthesis, and the optimal temperature range for development is 15–25°C. Moulds require the presence of moisture in the concrete surface. This can be from rainfall, but an alternative is prolonged periods when the ambient relative humidity is in excess of 80%.

The climate in the region where the mould staining occurred means that concrete surfaces can be damp or wet for prolonged periods of the year. The average relative humidity is over 80% from March to August, and this period can also bring significant rainfall. There can be lingering mist and fog on spring mornings. There is also some atmospheric pollution, which provides nutrients for mould growth in the form of particulate matter deposited on exposed surfaces. Wind direction dictates the orientation of surfaces on which deposition can occur, and rainfall can wash off nutrients from some surfaces.

Investigation

The principal objectives of the study were to find an appropriate method of cleaning the concrete surfaces and to recommend treatments to delay re-establishment of mould growth. Within these overall objectives there were several constraints:

- The bridges were to be kept fully open to traffic at all times.
- The appearance of the concrete surfaces was not to be altered (i.e. no surface coating).

In addition to these main constraints, because of the access difficulties, the number of operations had to be kept to a minimum to keep costs at reasonable levels.

Cleaning

The normal procedure used to remove mould growth from the surfaces of structures is to spray with biocide, allow a few days to pass to permit the biocide to take effect, and then to pressure wash. Biocides would also be used in the wash water to delay the onset of regrowth. As the main means of access to carry out the work on the bridge piers was cradles, the requirement to carry out two phases of cleaning separated by a few days would have needed more cradle movements or the provision of additional cradles.

In the circumstances it was appropriate to adopt a single operation strategy of power washing including a biocide. There would be a potential downside to this strategy in that it would be necessary to use higher pressures to remove organisms that had not been killed or weakened by prior application of biocide. This could lead to roughening of the surface and hence more rapid mould regrowth and greater difficulty of cleaning in the future.

Treatment of the surface

The features that promote mould growth have been discussed above. Of these, the only aspect that can be controlled relatively easily is the amount of moisture in the concrete surface. This control can be achieved by application of a surface treatment. Most products of this nature form films on the concrete surface and hence would have a radical effect on the appearance of the structure. This would have been contrary to one of the requirements of the investigation.

Hydrophobic impregnants such as silane and allied materials perform in a different way. They are absorbed into and bond with the underlying material. The materials line pores and create a hydrophobic effect without forming a surface film. Treated surfaces tend to dry out as the pores, although lined, remain open to the passage of air and water vapour. These types of material therefore appeared to be good candidates for application to the bridge piers.

Several materials are available on the market, from pure silanes to silanes in organic solvents and water-based silane/siloxanes. Pure silanes deliver concentrated reactive material to the surface, but they may not penetrate very deeply. Silanes in organic solvents materials may be able to penetrate deeper into the surface, but the concentration of active material is consequently lower. The concrete was of high quality and the low absorption was a further consideration in the choice of material. Silanes and siloxanes are normally applied in two coats. In this case, as access was difficult and costly, only one coat was to be applied.

Another influencing factor was that the material was to be applied from cradles, which meant that it was most efficient to work from the top down. Silanes and siloxanes are usually applied working from the bottom up. This makes the process easier to control as the material is being sprayed onto a dry surface rather than on a surface that has been dampened by material running down from the surfaces above, which have already been treated.

The views of local representatives of silane/siloxane manufacturers were consulted as to the most appropriate materials and method of application. The degree of local support for the products and local track records were also taken into consideration. Derivative silanes formulated as gels or pastes were chosen on the basis of this review. Materials formulated in this way had the following advantages:

- They are normally applied as a single-coat application, as the material persists on the surface for a short period and is absorbed into the surface during this time.

- They can be applied from the top down, as there is little run-down.
- The area treated is clearly visible while the material persists on the surface.
- There is a much lower likelihood of spray and run-down being deposited on vehicles using the bridges – this was an important consideration as the decks were to remain open to traffic during application.

Weather conditions during the treatment were also taken into account as part of the study, as it was understood that weather conditions could have a significant effect on the success of the process. As application was to be carried out from cradles, periods of high wind could result in excessive downtime and hence increased costs. It is also best to apply the materials to a dry concrete surface to increase absorption. It was recommended that the application should not proceed during the wet season and also that periods of the year when high wind speeds could be anticipated should be avoided.

Conclusions
The conclusions from the study were as follows:

- The concrete surfaces would be cleaned in a one-pass operation by jet-washing with biocide in the wash water.
- One coat of silane in gel form would be applied.
- The work would be carried out when suitable weather conditions prevailed (low wind speeds and low rainfall).

References

Alexander, M., B. Addis and J. Basson (1994) Case studies using a novel method to assess aggressiveness of waters to concrete. *ACI Materials Journal* **91**(2), 188–196.

Allen, R.T. (ed.) (1998) *Concrete in coastal structures.* Thomas Telford, London, ISBN 0 727726102, p. 145.

American Concrete Institute (2007a) Guide for concrete floor and slab construction, ACI 302.1R-04. In *ACI manual of concrete practice*, Part 2. American Concrete Institute, Farmington Hills, MI.

American Concrete Institute (2007b) Erosion of concrete in hydraulic structures, ACI 210R-93. In *ACI manual of concrete practice,* Part 1. American Concrete Institute, Farmington Hills, MI.

American Concrete Institute (2014) ACI 305. *Specification for hot weather concreting.* American Concrete Institute, Farmington Hills, MI.

Bahrain Society of Engineers (1985) *Proceedings of the first International Conference on Deterioration and Repair of Reinforced Concrete.* Bahrain Society of Engineers, Manama.

Bate, S.C.C. (1985) *High alumina cement concrete in existing building superstructures.* Building Research Establishment Report 235. HIS/BRE, Garston.

British Standards Institution (1972) CP 110. *Code of practice for the structural use of concrete.* British Standards Institution, London.

British Standards Institution (1987) BS8007. *Code of practice for design of concrete structures for retaining aqueous liquids* (withdrawn). BSI, London.

British Standards Institution (1990) BS 812-111. *Testing aggregates. Methods for determination of ten per cent fines value (TFV).* BSI, London.

British Standards Institution (2003) BS 8204-2. *Screeds, bases and in-situ floorings – Part 2: Concrete wearing surfaces – Code of practice.* BSI, London.

British Standards Institution (2006a) BS EN 1992-3. *Eurocode 2: Design of concrete structures. Part 3 – Liquid retaining and containment structures*. BSI, London.

British Standards Institution (2006b) BS 8500-1. *Concrete – Complementary British Standard to BS EN 206-1 – Part 1: Method of specifying and guidance for the specifier*. BSI, London.

British Standards Institution (2008) BS EN 12620: 2002. *Aggregates for concrete*, Amendment A1. BSI, London.

British Standards Institution (2013) BS EN 206. *Concrete. Specification, performance, production and conformity*. BSI, London.

Building Research Establishment (1991) *Sulfate and acid resistance of concrete in the ground*, Digest 363, ISBN 0 85125 500 0. BRE, Watford.

Building Research Establishment (2005) *Concrete in aggressive ground*. Special Digest 1, 3rd edn. BRE, Watford.

Buncefield Standards Task Group (2007) *Safety and environmental standards for fuel storage sites – Final Report*. BSTG.

Comité Européen du Béton (1992) *Durable concrete structures*, CEB Design Guide No. 183. Thomas Telford, London, ISBN 978-0-7277-1620-0, p. 75.

Concrete Society (2002) *Guide to the construction of reinforced concrete in the Arabian Peninsula*, Special Publication CS136/CIRIA Publication C577. Concrete Society, Camberley/CIRIA, London.

Concrete Society (2007) *Guidance on the use of macro-synthetic-fibre-reinforced concrete*, TR65. Concrete Society, Camberley.

Concrete Society (2008) *Guide to the design of concrete structures in the Arabian Peninsula*, Special Publication CS163. Concrete Society, Camberley.

Construction Industry Research and Information Association (1984) *The CIRIA guide to concrete construction in the Gulf region*, Special Publication 31. CIRIA, London.

De Muynck, W. *et al.* (2010) Microbial carbonate precipitation in construction materials: a review. *Ecological Engineering* **36**(2), 118–136.

Eglinton, M.S. (1987) *Concrete and its chemical behaviour*. Thomas Telford, London, ISBN 0 7277 0372 2, pp. 69–70.

Floyd, M. and D.E. Wimpenny (2003) Procedures for assessing thaumasite sulfate attack and adjacent ground conditions at buried concrete structures. First International Conference on Thaumasite in Cementitious Materials, Building Research Establishment, 19–21 June 2002. *Cement and Concrete Composites* **25**, 1077–1088.

Gould, M. (2003) The Ritchie system of reinforced concrete. *Transactions of the Newcomen Society* **73**, 275–291.

Hughes, P. (2013) A study into microbial growth within new marine concrete. *Concrete* **47**(1), 34–36.

Hughes, P. (2014) *An investigation into marine biofouling and its influence on the durability of concrete sea defences*, PhD thesis, University of Central Lancashire.

Hughes, P. *et al.* (2013a) Microbial degradation of synthetic fibre-reinforced marine concrete. *International Biodeterioration & Biodegradation* **30**, 1–4.

Hughes, P. *et al.* (2013b) Microbial degradation of synthetic fibre-reinforced marine concrete. *International Biodeterioration & Biodegradation* **30**, Section 2.3, p. 2.

Hughes, P. *et al.* (2013c) Briefing: Microscopic study into biodeterioration of joint sealant. *Construction Materials* **166**(5), 265–268.

Hughes, P. *et al.* (2013d) Microscopic examination of a new mechanism for accelerated degradation of synthetic fibre reinforced marine concrete. *Construction and Building Materials* **41**, 498–504.

Kay, E.A., P.G. Fookes and D.J. Pollock (1982) Deterioration related to chloride ingress. *Concrete in the Middle East*, Part 2. Eyre and Spottiswoode, London, Figure 22.

Kay, E.A., M.C. Mills, D.J. Pollock and T.J. Sharp (1979) *Concrete practice in the Middle East – Notes for guidance*. Halcrow International Partnership, Dubai.

M4L Programme. http://m4l.engineering.cf.ac.uk/9-uncategorised/69-home.html. Last accessed 30 September 2016.

Mason, P.A., P.R. Edwards and H.J. Amies (eds) (1997) *Design of containment systems for the prevention of water pollution from industrial incidents*, Report C164. CIRIA, London.

Page, C.L. and M.M. Page (eds) (2007) *Durability of concrete and cement composites*. Woodhead Publishing (Elsevier), Cambridge, p. 155.

Quillin, K. (2001) *Modelling degradation processes affecting concrete*, BR 434. Building Research Establishment, Watford, ISBN 1 86081 531 6, p. 22.

Rieder, K.A. (2007) *New concrete technology in construction, synthetic macro fibres*. Institute of Concrete Technology, Camberley, Annual Technical Symposium, pp. 1–14.

Tarada, F. and P.C. Robery (2014a) Containment for petroleum products – lessons learnt from Buncefield, UK. *Proceedings of the Institution of Civil Engineers, Civil Engineering* **167**(CE4), 167–175.

Tarada, F. and P.C. Robery (2014b) Containment for petroleum products – lessons learnt from Buncefield, UK. *Proceedings of the Institution of Civil Engineers, Civil Engineering* **167**(CE4), Figure 2, p. 169.

Thaumasite Expert Group (1999) *The thaumasite form of sulfate attack: risks, diagnosis, remedial works and guidance on new construction*. DETR, London, pp. 19–21.

UK Water Industry Research (2011) *Civil engineering specification for the water industry*, 7th edn. UKWIR, London.

Walton, I.L.W. (ed.) (2014) *Containment systems for the prevention of pollution –secondary, tertiary and other measures for industrial and commercial premises*, Report C736. CIRIA, London.

www.slp-precast.com/submat/images/precast/blackpool1.jpg; date last accessed 30 September 2016.

Chapter 3
Effects of typical extreme environments on concrete dams

Yan Xiang, Zhi-min Fu, Kai Zhang, Zhi-yuan Fang and Cheng-dong Liu

3.1 Introduction

Extreme events are the rare events that occur outside the statistical distribution in a particular period. They normally distribute on both sides of the statistical curve within a range of 5%, and have the characteristics 'severe', 'sudden' and so on. In the last 50 years, the frequency and intensity of China's extreme events have strengthened. Since the 1990s, a number of extreme events have occurred in China, causing significant losses to economic and social development. For example, Wenchuan earthquake in 2008 led to dangerous situations for 2,743 reservoirs and 822 hydropower stations because of the shock. However, specific details about the effects of extreme environments on dams are lacking. Jianyun *et al.* (2008) carried out a preliminary analysis of the impact of climate change on reservoir dams. Climate change may have important impacts on dam safety relating to: 1) changes in rainfall and runoff of the drainage basin due to climate change, thereby affecting the design for storms and flooding of the basin; 2) the possible exacerbation of the frequency, scope and extent of droughts, thereby affecting the guaranteed rate of water supply; 3) an increase in the intensity and frequency of storms, leading to geological disasters and increasing the impact of sediment on the safety and life of projects; 4) variability in the frequency and intensity of extreme hydroclimatic events, leading to extreme floods. Lee and You (2011) carried out a risk analysis of the impact on reservoirs of extreme hydrologic events due to climate change. Their research showed that, in the context of global climate change, the frequency and intensity of floods, droughts, typhoons, earthquakes and other natural disasters will increase, and the impact on reservoir dams will be more serious. The effect of two typical extreme environments, climate fluctuations and earthquakes, on concrete dams will be studied in this chapter.

3.2 Analysis of the effect of the Wenchuan earthquake on a concrete dam

On 12 May 2008, an earthquake of magnitude 8.0 on the Richter scale occurred at latitude 31.0° and longitude 103.4° in Wenchuan, Sichuan Province, China, with an epicentral intensity of 11 degrees and a rupture length of 300 km, caused by a seismogenic fault, and with a duration of 100 s. The earthquake generated strong destructibility, affected wide areas and led to serious disaster in the region, with a serious impact on dams and other infrastructure. There are 1,997 reservoirs and more than 800 hydropower stations in Sichuan Province, which suffered various degrees of shock

Figure 3.1 Shapai hydropower station after the earthquake.

Figure 3.2 Full view of Tongkou hydropower station.

damage. Taking two concrete dams within the range of the earthquake, Shapai hydropower station (Figure 3.1) and Tongkou hydropower station (Figure 3.2) as examples, the impact of earthquakes on concrete dams was investigated and analysed.

3.2.1 Investigation of earthquake damage at Shapai hydropower station

The Shapai hydropower station project is located in Wenchuan County, Sichuan Province, 36 km from the epicentre of the earthquake (Figure 3.3), 8 km from the fault zone of Longmen Mountain and 29 km from the central fault of Longmen Mountain. It consists of a roller-compacted concrete (RCC) arch dam, two spillway tunnels, an intake tunnel and a power house on the right bank. The project has a total storage

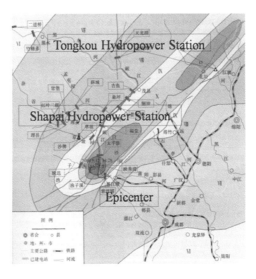

Figure 3.3 Distribution map of Wenchuan earthquake intensity near the hydropower station.

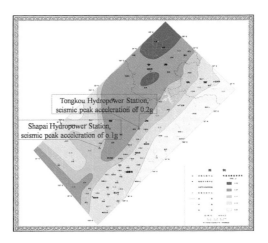

Figure 3.4 Distribution map of Wenchuan earthquake peak acceleration near the hydropower station.

capacity of 0.18×10^8 m³, a power plant installed capacity of 36 MW, a maximum height of 132 m, basic seismic intensity of VII, dam fortification intensity of VII and level peak acceleration of the corresponding bedrock of 0.141 g (Figure 3.4). Before the earthquake, the water level of the reservoir was normal and the power plant was operating normally.

3.2.1.1 Water-retaining structure
After the earthquake, the dam was as good as whole (Figures 3.1, 3.5 and 3.6). The dam structure, foundation and abutment showed no abnormal changes. However, a

Figure 3.5 The upstream face of the dam after the earthquake.

Figure 3.6 The downstream face of the dam after the earthquake.

rock fall occurred on part of the superficial rock of the resisting slope, and the top of a transverse joint at the right side of the dam cracked. There were several cracks after the earthquake between the filler wall of the elevator shaft operating room of the dam crest and concrete frame, and there was partial shedding of decorative material. The earthquake damage level was slight.

3.2.1.2 Discharge structures
The main buildings of the two spillway tunnels were in good condition. The gates opened and closed normally. A small amount of collapse happened on the superficial near outlet slope. The earthquake damage level was slight.

3.2.1.3 Power generation structures
The structure of the main and auxiliary powerhouses along the side of the road and roof structures were damaged seriously by landslide and rolling stones, and the

Figure 3.7 Seriously damaged powerhouse roof structures.

Figure 3.8 Collapsed slope of the surge shaft.

tailraces and lock chamber structure were slightly damaged (Figure 3.7). The main and auxiliary powerhouses and plant area were flooded. The intake gate of the diversion tunnel opened and closed normally, but the bent frame column of the hoist had shear cracks. The slope of the surge shaft and many sections of the entry road collapsed (Figure 3.8). The plant area was seriously damaged. The earthquake damage to the diversion tunnels was slight, and that of the powerhouse was relatively serious.

3.2.1.4 Electromechanical equipments and metal structures
The bridge crane of the main powerhouse was hit by huge stones and collapsed. The left bellow of the crossing bridge of the steel diversion pipe was fractured, and subjected to high-pressure water jets, and the generators and other equipments were

Figure 3.9 The flooded generators.

Figure 3.10 The damaged switching station.

flooded (Figure 3.9). Some equipments of outgoing line yard and switching station were damaged (Figure 3.10). The earthquake damage level was relatively serious.

3.2.1.5 Others

The engineering slope of the dam abutment resisting rock was good, but the local natural slopes of the two sides above the crest elevation collapsed (Figure 3.11). The slope near the reservoir bank was in good condition, and its earthquake damage

Figure 3.11 The collapsed local natural slope.

Figure 3.12 External transportation was seriously collapsed.

level was slight. The powerhouse and adjoining scree slopes collapsed, causing serious earthquake damage (Figure 3.12). The earthquake damage level was relatively serious.

3.2.2 Investigation of earthquake damage at Tongkou hydropower station

Tongkou station is located in Beichuan County, Sichuan Province, 2 km upstream of Tongkou town, with a total storage capacity of 3.61×10^6 m³, an installed capacity of 45 MW and a maximum height of 71.5 m. It mainly comprises an RCC gravity dam, a water diversion system with right bank, main and auxiliary ground powerhouses, a booster station, and so on. The dam has an overflow dam section and a non-overflow dam section. The overflow dam section consists of a five-hole surface spillway and a

Figure 3.13 The damaged railings of the dam crest.

one-hole bottom flushing outlet. The bottom flushing outlet is used for sand washing, flood discharge and reservoir emptying. The power station intake is arranged on the right bank, connecting with the main powerhouse through a diversion tunnel and penstock.

Tongkou hydropower station is located in the northern section of Longmen Mountain middle fold belt, on a fault block between Yingxiu fault and Jiangyou-Guan County fault (Figures 3.3 and 3.4). The station region is a substability.

3.2.2.1 Water-retaining structure
Local fractures occurred at the settlement joint of the traffic bridge surface and caused damage to the railings (Figure 3.13), while the water-retaining structure showed no abnormal changes overall. The expansion joints between the #2 and #3 dam section were cracked to a width of about 1 cm. Seepage as well as leakage of the foundation and drainage gallery was larger after the earthquake, but the dam remained safely under control. In summary, earthquake damage to this dam was slight.

3.2.2.2 Water conveyance and power generation structures
The importing slope of the water conveyance system and the main structure of the powerhouse showed no abnormal changes, but the surface of the local structural joint of the main powerhouse and the local walls of the distribution substation were cracked, and the filler wall was damaged severely (Figure 3.14). Moreover, the channels after the slopes on both banks of the tailwater channel contained debris. Earthquake damage level was slight.

3.2.2.3 Electromechanical equipment and metal structures
The gates and hoisting equipment showed no abnormal changes. The water seal of #4 floodgate was damaged and the gate was offset 5 cm to the right. The earthquake damage level was slight.

Figure 3.14 The filler wall was severely damaged.

3.2.2.4 Other structures

There were local cracks in the overburden layer slope of the right bank above the crest elevation, with a maximum width of 8–10 cm. The slope after the powerhouse showed no abnormal changes. The earthquake damage level was slight.

Flow within the powerhouses showed no abnormal changes, but movement of part the road concrete pavement occurred. Parts of the safety monitoring facilities had loosened screws. Earthquake damage level belonged to slight.

3.2.3 Characteristics of seismic damage on the concrete dams

3.2.3.1 Large concrete structures

Earthquake damage to the large concrete structures was rare. The large concrete structures forming the dam, water conveyance structures and so on showed no abnormal changes. The concrete dam project withstood the test of a strong earthquake far beyond the fortification standard (*Specification for seismic design of hydraulic structures* (SL 203-97),1997; *Code for seismic design of hydraulic structures of hydropower project* (NB 35047-2015), 2015), testifying that the current seismic design theory and method of the standard can ensure safety and reliability of such a project. Additionally, seismic fortification for the important water-retaining structures further ensure aseismic safety of dam projects based on current seismic fortification methods in China. In summary, large concrete dams have a good anti-seismic capability.

3.2.3.2 Abutments and slopes

The abutments and engineering slopes were stable overall and showed little earthquake damage. The abutments on both sides were reinforced by prestressed cables and the engineering slopes were reinforced by systematic bolt-shotcrete supports, and were not damaged by the earthquake. The natural slopes near the engineering slopes appeared to have collapsed and showed cracks, indicating that the artificial slope reinforced by engineering measures had better stability. The abutment stability and slope stability effectively ensured the safety of the dam. Therefore, high attention should be paid in earthquake zones to the stability of the dam abutment and engineering of the slope environment, and necessary engineering measures should be undertaken.

3.2.3.3 Underground structures
Earthquake damage to the underground structures was slight; they have strong earthquake-resistant capability, including water diversion and the power generation tunnel. At Shapai hydropower station, relatively serious earthquake damage occurred at the import and export of underground structures, with most of the damage being caused by slope collapse. Cracks at Tongkou hydropower station were only found at the opening of the slope.

3.2.3.4 Secondary impacts
Secondary effects from earthquakes have a great impact on buildings. Buildings and equipment are often damaged by landslides, damming, flooding and other disasters after earthquakes. The powerhouses at Shapai hydropower station suffered heavy earthquake damage, mainly due to mountain collapse. The earthquake did not cause direct and obvious damage to the hydraulic structures, but slope collapse near the powerhouses caused by the earthquake led to a large area of landslide and rolling of huge rocks, which damaged the bridge crane. The electromechanical equipment in the outgoing line yard of the powerhouse switching stations was hit and damaged to varying degrees by flying stones.

3.2.3.5 Auxiliary buildings
The auxiliary buildings, consisting of beams, columns and walls, suffered relatively serious earthquake damage. The implication of earthquakes on important subsidiary structures related to engineering safety in the earthquake zone should be paid more attention, in particular in relation to bending of the headstock gear of the flood discharge structures. If necessary, seismic design safety standards should be improved to ensure that sluice gates can be opened smoothly in emergency situations.

3.3 Effect of extreme temperature change on concrete dams
Xixi reservoir is located near Shadi village on Daxi River, which is a tributary of Baixi River in Ninghai County, Zhejiang Province. It is 16 km away from the County and 8 km downstream of Huangtan reservoir. The main project for the reservoir began in October 2003, began to impound in July 2005, and was completed in August 2006. The reservoir is a multi-year regulating reservoir and a comprehensively utilised medium-sized hydraulic engineering with the main objectives of flood control and water supply, as well as irrigation and electricity generation. The dam is an RCC gravity dam with a total of 13 dam monoliths and a maximum height of 71.0 m. It includes ten non-spillway dam monoliths, a diversion dam monolith and two spillway dam monoliths. The geographical position and downstream elevation of the dam are shown in Figure 3.15. It has a normal water level of 147.00 m, a design flood level of

152.02 m and a check flood level of 152.45 m. The heavy rains of No. 16 Typhoon 'Rosa' saw the reservoir reach its highest level (151.29 m) since its construction, on 7 October 2007. In order to measure the uplift pressure on the dam, 11 uplift pressure test holes were arranged in foundation corridors #3 to #12 of the dam monolith. The depth of the holes was 1.0 m below the base surface. Meanwhile, to monitor the temperature in the bedrock, bedrock thermometers were also laid.

3.3.1 Analysis of observational seepage data for Xixi reservoir

The period over which monitoring data were analysed extended from 27 June 2007 to 22 November 2010. Prior to 12 April 2010, UP4 and UP5 uplift pressure coefficients were below 0.042 and 0.065 and the amplitude was small. From 13 April 2010 to 28 April 2010, UP4 and UP5 uplift pressure coefficients surged to their historical maxima. The values for the other dam monoliths did not show any obvious abnormal increases. In order to analyse the sudden increases in the coefficients, plots of uplift pressure coefficient and change in upstream water level (Figure 3.16) and plots of uplift pressure coefficient and change in bedrock temperature (Figure 3.17) were drawn. Changes in the various environmental factors when the reservoir was operating with high water levels are shown in Table 3.1.

Figures 3.16 and 3.17 show that the water level of 147.00 m on 7 October 2007 and 28 October 2007, the highest water level of 151.29 m during the period, which lasted 22 days. The air temperature changed from 24.3°C to 15.3°C, while the water

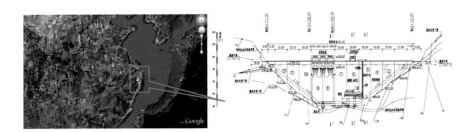

Figure 3.15 Geographical position and downstream elevation of Xixi reservoir dam.

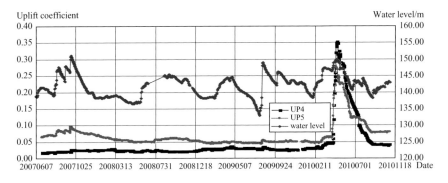

Figure 3.16 Plot of UP4 and UP5 uplift pressure coefficients and water level.

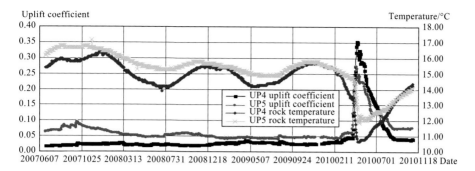

Figure 3.17 Plot of UP4 and UP5 uplift pressure coefficients and corresponding bedrock temperature.

Table 3.1 Calculated uplift pressure coefficients for different temperatures and three different water levels.

Parameter	Serial number	H_0	$H - H_0$	$T - T_0$	$(H - H_0)^2$	$(T - T_0)^2$	α
Normal water level	1	147.00	0.18	−1	0.0324	1	0.124747
	2	147.00	0.18	−2	0.0324	4	0.201982
	3	147.00	0.18	−3	0.0324	9	0.275731
	4	147.00	0.18	−4	0.0324	16	0.345993
	5	147.00	0.18	−5	0.0324	25	0.412768
	6	147.00	0.18	−6	0.0324	36	0.476057
Design flood level	1	152.02	5.2	−1	27.04	1	0.341594
	2	152.02	5.2	−2	27.04	4	0.418829
	3	152.02	5.2	−3	27.04	9	0.492578
	4	152.02	5.2	−4	27.04	16	0.56284
	5	152.02	5.2	−5	27.04	25	0.629615
	6	152.02	5.2	−6	27.04	36	0.692903
Check flood level	1	152.45	5.63	−1	31.6969	1	0.366626
	2	152.45	5.63	−2	31.6969	4	0.443862
	3	152.45	5.63	−3	31.6969	9	0.51761
	4	152.45	5.63	−4	31.6969	16	0.587872
	5	152.45	5.63	−5	31.6969	25	0.654647
	6	152.45	5.63	−6	31.6969	36	0.717936

temperature at the bedrock surface maintained a temperature of about 21.9°C. UP4 and UP5 uplift pressure coefficients were near 0.025 and 0.096, without obvious change. From 10 March 2010 to 1 May 2010, the water level rose from 147.00 m to 149.86 m, and for 51 days sustained this high water level. During this period the rainfall was 51 mm and 36.6 mm on 11 April and 16 April, respectively, the air temperature ranged from −0.4°C to 18.6°C (including the temperature plummeting from 17.4°C to 7°C on 1 April), the water temperature of the bedrock surface was maintained at 10.2°C, and UP4

57

and UP5 uplift pressure coefficients were 0.353 and 0.276. Therefore, the duration of the high water level (147.00 m) experienced in July 2007 was short, and the temperatures were relative high. From April to May 2010, the high water level (up to 148.90 m) lasted longer and the temperature of the bedrock was lower. The anomaly where the uplift pressure increased for UP4 and UP5 occurred on 15 April 2010, and this can be understood as follows: at low temperatures the bedrock fracture opened, weak flow channels of bedrock opened, and low-temperature water passed through the flow channels, leading to a reduction in the temperature of the cracks, which opened further. With drawdown of the water level and a rise in temperature, the uplift pressure gradually reduced. On 12 November 2010, the uplift pressure coefficients were 0.041 and 0.079, respectively.

3.3.2 Calculation and analysis of the intensity and stability of the gravity dam under uplift pressure

Through analysis of the observed seepage data, it was found qualitatively that low temperature and an increase in the reservoir water level lead to an increase in the uplift pressure of the dam foundations. To further analyse the relationship between uplift pressure coefficients and temperature change and water level, a statistical regression model of the uplift pressure coefficients was set up to quantitatively analyse the effect of low temperature and water level on dam safety.

From experience, the function relating pressure coefficient α and temperature $(T - T_0)$ and water level changes $(H - H_0)$ has been constructed:

$$\alpha = A_1(H - H_0) + A_2(T - T_0) + B_1(H - H_0)^2 + B_2(T - T_0)^2 + C \qquad (3.1)$$

where H is water level in metres, H_0 is initial water level in metres, T is temperature in °C, T_0 is initial temperature in °C, α is the uplift pressure coefficient, and A_1, A_2, B_1, B_2 and C are regression coefficients.

By using observed seepage data and the regression calculation, the regression coefficients (A_1, A_2, B_1, B_2 and C) were obtained (Figure 3.18), giving the following final function:

$$\alpha = 0.0284(H - H_0) - 0.0825(T - T_0) + 0.0028(H - H_0)^2$$
$$- 0.0017(T - T_0)^2 + 0.0388 \qquad (3.2)$$

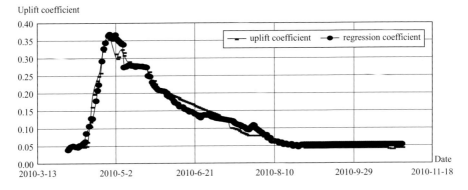

Figure 3.18 Measured uplift coefficient and regressed uplift coefficient.

58

Table 3.2 Calculated results for dam stability and stress at three different water levels.

Characteristic water level Temperature difference (°C)	Normal water level			Design flood level			Check flood level		
	α_1	K_1'	σ_1	α_2	K_2'	σ_2	α_3	K_3'	σ_3
−2	0.202	3.993	0.310	0.419	3.290	0.080	0.444	3.264	0.059
−4	0.346	3.809	0.250	0.563	3.106	0.027	0.588	3.081	0.006
−6	0.476	3.643	0.195	0.693	2.940	−0.020	0.718	2.914	−0.039

α_i, K_i' and σ_i (i = 1, 2, 3) represent the uplift pressure coefficient, sliding stability factor and dam heel stress, respectively.

According to this derived function, values of α_i (i = 1, 2, 3) were calculated by a mechanics of materials method for the normal water level, design flood level and check flood level at different temperatures. For the three characteristic water levels (normal, design flood and check flood), for temperature differences ($T - T_0$) under six temperature conditions (1°C, 2°C, 3°C, 4°C, 5°C and 6°C, respectively), the corresponding uplift pressure coefficients were calculated by applying equation 3.2. The results are shown in Table 3.1.

According to an on-site test of the volume density of the RRC dam, the average for concrete with two grades of aggregate is 23.31 kN/m³ and for concrete with three grades of aggregate is 23.35 kN/m³, and the shear strength indicators for the base surface are $f' = 0.9$ and $c' = 0.8$ MPa. The uplift pressure coefficients were calculated and analysed for the normal water level at different temperatures. The calculated results for dam stability and stress are provided in Table 3.2. These show that, under normal-water conditions the temperature T_0 would be set as −2°C, −4°C and −6°C, the anti-sliding stability factor K' and the stress of the dam heel meet specifications. Under design flood conditions and check flood conditions, when temperatures are −4°C and −2°C, requirements are still met. As the temperature continues to decline and the temperature difference reaches −6°C, tensile stress appears for design flood level and check flood level conditions, thus not meeting stress requirements. As the temperature continues to decrease (the temperature difference increases), an inflection point appears, which means the requirements for dam stress and stability are not met. To ensure safe and reliable operation of the dam, the threshold where the strength and stability of the dam change must be determined.

3.4 Behaviour analysis of the safe operation of Shenwo reservoir in an extreme cold environment

3.4.1 Influence of a cold climate on safe operation of concrete dams

Concrete buildings are inevitably damaged by alternating drought and wet, cold and hot, and freezing and thawing, conditions that are found in the environment around Shenwo reservoir. The utilisation and lifetime of such buildings will therefore be adversely affected. Indeed, freeze–thaw-induced spalling is found in 22% of concrete dams and in 21% of the associated small- and medium-sized reinforced concrete structures (e.g. sluices). The influences of a cold climate on the safe operation of a concrete dam may be summarised as follows:

3.4.1.1 Effects of low temperature on the hardening and curing of concrete

Temperature has a direct effect on cement hydration, and the rate of hydration is directly proportional to temperature. As a general rule, 20°C is regarded as a limit to the normal curing process. The rate of hydration is expected to increase (or decrease) if the temperature is higher (or lower) than 20°C. The hydration process is supposed to stop when the temperature is lower than 0°C. For example, hardening of concrete at 4°C takes twice the time than at 20°C. This is because the free water contained in the concrete freezes. As a result, hydration stops. The freezing of fresh concrete reduces its long-term strength and causes cracking. Because the water in fresh concrete causes saturation, once the concrete is frozen the coagulating effect between the cement and any admixtures diminishes. In addition, free water on the aggregates is likely to turn to a thin film of ice and free water becomes separated from the grout, which leads to a loss of strength.

3.4.1.2 Influence of low temperature on the mechanical properties of concrete

The temperature in the concrete is determined by the heat energy required for hydration. Because the temperature in the concrete differs from that of the external environment, a lower ambient temperature is expected to reduce the temperature of the concrete through heat exchange. The mechanical parameters of the concrete are changed as follows:

- *Compressive strength*. Cracks expand and connect with each other. Relevant studies demonstrate that cracks are first generated on the interface between the cement gel and the aggregate in a concrete under compression. With increasing compressive stress on the concrete, these cracks extend to the cement gel. Meanwhile, the cracks in the interface begin to expand, which leads to a loss of strength and damage to the concrete. The compressive strength increases over time: with lower (and even negative) temperatures, the compressive strength is only slightly enhanced.
- *Elastic modulus*. This is an important mechanical property of any concrete and reflects the relationship between the stress suffered by and the strain generated in the concrete. Meanwhile, it is necessary when calculating structural deformation, crack development and thermal stress in a mass concrete. The strength and elastic modulus are influenced by the porosities of the various components in a concrete. Studies have indicated that the elastic modulus increases with age of the concrete, but this is a slow-growth trend.
- *Tensile strength*: The low tensile strength of concrete causes cracking. This affects its appearance and durability, and even jeopardises the safety and normal utilisation of concrete structures to a certain extent. The tensile strength of concretes cured for 28 days at −30°C is 118% that of concrete undergoing standard curing for the same period at 20°C. The tensile strength of concrete cured for 28 days at −20°C and −10°C is 109% and 105%, respectively, that of

concrete cured for 28 days at 20°C. Therefore, the tensile strength of concrete is increased by a reduction in temperature.

3.4.1.3 Influence of the freezing process on safe operation of concrete dams

The freezing of concrete is a long-term, complicated, physical process, the action mechanism of which has not yet been determined. The freezing of concrete is affected by several factors, including saturation of the concrete, its pore structure, the chemical composition of the pore water, freezing rate, temperature cycle, and so on.

When it comes to durability, the repeated freezing of concrete causes its gradual destruction (freeze–thaw-induced damage). The freeze–thaw-induced destruction is actually a process during which the structure of the hydration products is converted from a compact state to a loose state, accompanied by cracking. However, the components of the hydration products remain unchanged and the process is largely physical.

From the perspective of structural distortion, when concrete is affected by freezing, apart from the shrinkage caused by the temperature reduction, there is some expansion as the freezing front gradually moves inwards. Meanwhile, as the freezing front migrates inwards, shrinkage is generated in addition to the expansion induced by the temperature increase. The repeated freezing process is also likely to generate a residual swelling deformation in the concrete.

3.4.2 Introduction to the project

Shenwo reservoir is located on the main stream of the Taizi River, Liaoyang, Liaoning Province, China. The controlled catchment area of the dam and the annual average runoff are 6,175 km² and 2.45 billion m³, respectively. It is a large-scale hydro-junction mainly designed to control flooding but also used to supply irrigation and industrial water, and generate power while supplying water. In this reservoir, the standards for design and check floods have 300 and 10,000 year recurrence intervals, respectively, and the corresponding water levels are 101.8 m and 102 m. The total storage volume is 0.791 billion m³, and the layout is as shown in Figure 3.19.

The dam is a concrete gravity dam, and the crest elevation, maximum dam height and total length of the dam crest are 103.5 m, 50.3 m and 532 m, respectively. This dam is composed of three parts, including retaining, overflow and power station dam

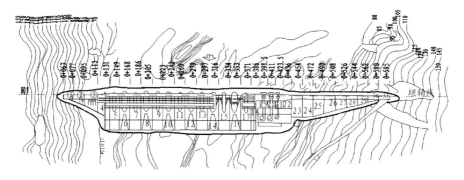

Figure 3.19 Layout of Shenwo reservoir dam.

Figure 3.20 Shenwo dam: (a) upstream and (b) downstream.

Table 3.3	Monthly average temperatures at the dam site from 2005 to 2010 (in °C).						
Month	2005	2006	2007	2008	2009	2010	Average
1	−12.90	−11.80	−9.50	−10.30	−11.00	−11.50	−11.17
2	−13.30	−8.26	−2.20	−7.40	−7.50	−7.96	−7.77
3	−1.90	−0.26	0.38	4.16	−0.59	−3.20	−0.24
4	9.90	7.90	9.25	12.00	11.95	6.33	9.56
5	14.40	17.20	18.10	16.24	19.50	16.30	16.96
6	21.10	22.20	23.60	21.85	21.60	23.51	22.31
7	24.50	24.50	24.60	25.80	24.60	25.02	24.84
8	23.20	25.10	23.20	25.00	25.50	23.07	24.18
9	18.00	18.40	19.80	18.00	18.60	18.05	18.48
10	9.95	11.90	10.30	12.20	11.20	10.04	10.93
11	0.54	2.00	2.95	1.89	−1.37	1.55	1.26
12	−12.00	−5.50	−5.70	−5.81	−8.90	−8.20	−7.69

sections, containing 31 dam sections in total. Dam sections #1–3 (right-hand side) and #22–31 (left-hand side) belong to the retaining part, with a total length of 217.3 m. Dam sections #4–18 and #19–21 form the overflow and power station parts, respectively, with lengths of 274.2 m and 40.5 m, respectively. The upstream and downstream faces of the dam are shown in Figure 3.20.

The annual average temperature at Shenwo reservoir is less than 10°C, and the average temperature in January is −11.2°C, with the minimum temperature reaching −35°C. The annual difference in the monthly average temperature exceeds 35°C, and the winter lasts as long as five months (from November to March). The monthly average temperatures at the dam site from 2005 to 2010 are provided in Table 3.3.

Limited by historical factors, the concrete in Shenwo dam is endowed with poor properties arising from the practices adopted during construction. Core samples demonstrated that for $R_{28} = 200^\#$, the maximum and minimum compressive strengths of the 75 core samples, were 54.5 MPa and 9.8 MPa, respectively, at 150 days, showing significant variability. For the concrete $R_{28} = 200^\#D100$, the rates of frost resistance were

41.2% and 72% in 1971 and 1972. In 2012, a safety inspection was carried out by Nanjing Hydraulic Research Institute. During the inspection, one group of specimens was taken from dam sections, including #3 and #29 on the upstream face, and two groups were taken from dam section #21 on the downstream face for indoor frost resistance testing. The experimental results demonstrated that the frost resistance grade of the sampled intact concrete failed to reach design requirements. Shenwo dam has suffered severe freeze–thaw-induced damage since its construction. The damage suffered by the retaining and power station sections of the dam in 2012 can be summarised as follows:

- *Retaining dam section*: A group of core samples were drilled from 0.1 m above the construction joint, with the stake number and elevation being 0+445 and 90 m, respectively. The depth of the freeze–thaw-induced damage was then observed. In the core samples, the upper part (8 cm thick) was an insulating layer made of polystyrene boards. Below that layer, broken concrete was found from 8 to 25 cm, showing C-type freeze–thaw denudation.
- *Power station dam section*: Intact core samples were taken from 0+423 and 68 m from dam section #21 on the downstream face. The depth of the freeze–thaw-induced damage in these core samples was 22 mm (B-type freeze–thaw denudation). In dam section #20, the concrete in the construction joint at an elevation of 80 m was severely corroded by the freeze–thaw process. Meanwhile, the concrete on the surface layer was broken and spalled, with a spallation depth of 10–34 mm. The depth of the damaged concrete on the surface layer was ~60–265 mm (C-type freeze–thaw denudation).

3.4.3 Analysis of dam deformation under the influence of a cold environment

3.4.3.1 Horizontal displacement of the dam crest

For the 31 dam sections (except for dam sections #1–3, #19, #22 and #29–31) each deformation mark point was set on the dam crest near the upstream surface of each section, giving 25 measurement points in total. Horizontal displacements downstream and upstream are taken to be positive and negative, respectively. The horizontal displacements of the dam crest from 1 January 1976 to 31 January 2011 are shown in Figure 3.21.

The variation in the horizontal displacement of the dam crest was negatively correlated with temperature. The lower the temperature, the larger the horizontal displacement (and vice versa). As a general rule, displacement (downstream) was greater from December each year to March of the following year, while that along the upstream face was more significant from June to October. Both sides of the dam sections were significantly affected by temperature. The concrete in the downstream dam face is more sensitive to temperature variations than that in the upstream dam face. According to the principle of thermal expansion, the concrete in the downstream dam face presents higher temperatures than the concrete in the upstream face as the temperature rises, leading to displacement of the dam towards the upstream side. When the temperature drops, the temperature of the concrete in the downstream face is lower than in the upstream face, resulting in displacement of the dam towards the downstream side.

In addition, the majority of the dam crest data showed a tendency to horizontal upstream displacement, especially in the retaining dam sections on both sides. This is

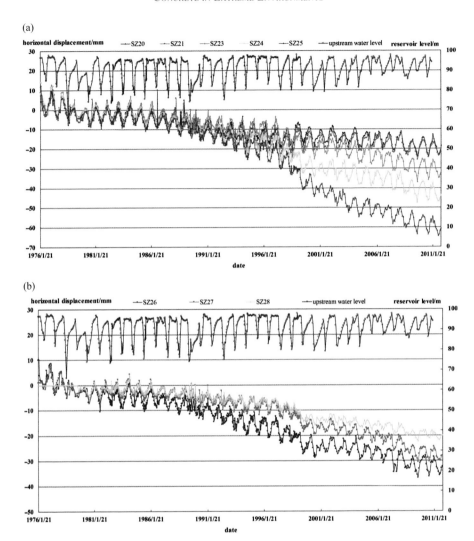

Figure 3.21 Measured horizontal displacement of the dam crest including (a) SZ19–SZ25 and (b) SZ26–SZ31.

contrary to the general behaviour seen elsewhere: concrete gravity dams tend to move downstream under the influence of the water load. Preliminary analysis was carried out to investigate this. Shenwo reservoir was built during the Cultural Revolution, and the concrete in the dam was poorly cast. Furthermore, inadequate temperature control measures were taken. The reservoir is located in the cold area of northern China, with large temperature differences between winter (the temperature is generally −20°C) and summer, and between day and night, and the phenomenon of freeze–thaw-induced denudation is found in the negative-temperature regions of the dam. As temperature deformation is elastic, freeze–thaw deformation leads to the generation of unrecoverable residual deformation each year. As a result, the negative-temperature region of the concrete in the downstream face of the dam is far greater than that in the upstream face. The annually accumulated residual deformation in the downstream face of the dam is thus greater than that in the upstream face, resulting in the dam crest moving upstream year by year. This is coincident with the phenomenon found in the downstream face of the dam, where large areas of cracks and denudation appeared in the concrete.

3.4.3.2 Vertical displacement of the dam

Twenty-eight benchmarks were arranged along the dam crest (including dam sections #2–29) from left to right, with the corresponding monitoring numbers CX02–CX29. The sinking and rising of the dam are represented by positive and negative values, respectively. The vertical displacements of the dam crest from 3 April 1982 to 20 October 2011 are shown in Figure 3.22.

(a)

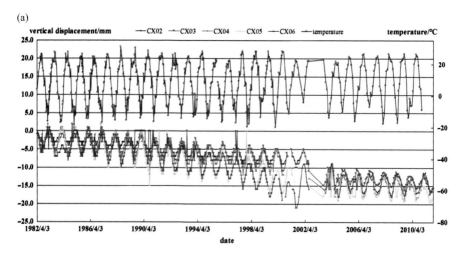

(b)

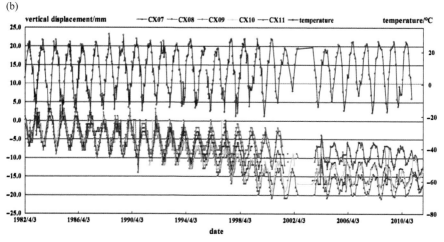

Figure 3.22 Vertical displacement of the dam crest for (a) CX04–CX06 and (b) CX07–CX11.

65

The measured vertical displacements presented a significant annual variation, and were mainly influenced by temperature variation. When the temperature rose, the crest lifted and the measured vertical displacement decreased, and vice versa. Temperature exerted a hysteretic influence on the vertical displacement. The uplift was greatest from July to October of each year, particularly in August, while the sinking was greatest from December of each year to March of the next year.

The vertical displacement at most points gradually decreased; that is, the dam tended to be uplifted year by year. The foundation of the dam sits on Presinian metamorphic rocks, most of which comprise hard and compact leptynites such as quartzs and biotites, with local areas having distributed schists including biotites and chlorites in strips and quartz veinlets. The strata are old in this area and the dam has been in operation for 40 years. Accordingly it is not possible that the gradual rise of the dam crest is a result of lifting of the dam foundation as a consequence of the geological conditions. Instead, it is primarily caused by the influence of frost heave. Frost heave leads to the generation of irrecoverable residual deformation in the dam each year, the accumulation of which is shown as a time-dependent deformation, resulting in the rise of the crest year by year. In addition, the freeze–thaw action is obvious within the cracks in the retaining dam sections of both sides, where considerable numbers of cracks can be found. This is more remarkable in the power station dam at an elevation of 78–89 m, where horizontal seepage cracks are found, and in the non-overflowing dam sections in the left-hand side. As a result, the uplift of the crest in these dam sections is more significant than in the overflow dam section.

3.4.4 Analysis of dam cracks under the influence of a cold environment

The concrete in the dam at Shenwo reservoir is of poor pouring quality, and it was affected by inadequate temperature control and low frost and crack resistance, as well as the widespread existences of potholes and cold joints. Cracks were found in the dam in August 1971 for the first time, and 210 cracked segments were found in the same year. The dam has been severely weathered and corroded after being operated for many years, with the number of cracks gradually increasing. The data for crack observations over the years are summarised in Table 3.4. In 2012 the Reservoir Administration Bureau carried out a field survey to study cracks in the sluice piers, and found that the number, length and width of the cracks in the dam are increasing year by year, and the problem of water seepage is becoming increasingly severe. The cracks in some typical sections are shown in Figure 3.23.

To study the crack development in a more detailed way, core drilling was performed on the cracks with serial number 29S-1 (denoting dam section #29 on the upstream face). The crack at 0+478.0 and 72 m in dam section #25 and that at 0+420.0 and 67 m in dam section #21 on the downstream face were also tested. The test results indicated that the crack depth in the core samples collected from dam section #21 exceeded 45 cm, and the crack width was 1.7 mm. With increasing depth, the width did not decrease. Meanwhile, core samples obtained from dam section #29 were broken along the crack, which was more than 43 cm long and 2 mm wide. Similarly, the width did not decrease with increasing crack depth. The strike of the crack in the

Table 3.4	Number of cracks found in the field survey.								
Serial number	1	2	3	4	5	6	7	8	9	10
Year	1971.12	1973.3	1973.9	1974.4	1975.3	1981.5	1983.8	1986.6	1997	2012
Total number of cracks	210	343	369	455	466	641	688	812	913	1243
Number of visible cracks		52		62	104	104	74			

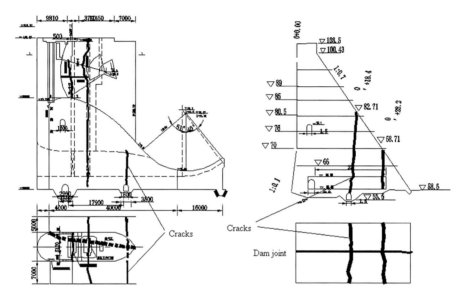

Figure 3.23 Cracks in some typical sections of the dam.

sample collected from dam section #25 was not perpendicular to the surface of the dam. The crack, with a width of 2.2 mm, extended out of the core sample at a depth of 15 cm. Similarly, the width of the cracks remained unchanged, or tended to increase, with depth.

3.4.4.1 Qualitative analysis of the cause of the cracks

The reasons for the generation of cracking may be summarised as follows:

- While constructing the dam, the engineers did not have sufficient knowledge of the distribution of thermal stress in such concrete dams in cold regions. They also failed to control the temperature and subsequent cracking. Therefore, the temperatures in the dam were high, causing significant thermal stress. According to the design requirements, the maximum temperature of the concrete in the dam foundation must be no more than 28–32°C (that is, the ambient

67

temperature of 7°C plus the allowable temperature difference of 21–25°C). However, according to data measured during construction, the maximum temperature of the concrete in the foundation was generally between 40°C and 45°C, with minimum and maximum temperatures of 35.6°C and 49.4°C, respectively. The temperature was far greater than the design standard. As a result, cracks were found in this dam for the first time in August 1971. The fractures generated on the dam surface during the construction period were mainly caused by the sudden drop in temperature and the large temperature difference between the inside and outside of the concrete. To be specific, the following four factors inhibited temperature control. The first factor was the heat of hydration of the concrete. The cement content in this dam is high: concrete 200# accounted for 90% of the total amount used. Meanwhile, the average cement content per cubic metre was 220 kg. The second factor was over-deep concrete lifts: pours more than 1.5 m and 5 m thick accounted for 90% and 37% of the total pours placed, respectively. The third factor was long pours. The method of pouring without longitudinal joints was adopted in the foundation to allow concrete strips up to 40 m long to be poured at once. Due to deterioration of the constraint conditions in the foundation, high shrinkage stresses were generated during cooling. Finally, summer pours were too hot (18–27.5°C), and these accounted for a third of all pours, leading to an increased maximum temperature in the concrete.

• Some concrete was non-uniform (coefficient of variation, $C_v > 0.2$), while some concrete failed to reach the design standard. All of these factors reduced the crack resistance of the concrete.

• Coldspells were frequent during the construction period, leading to a large temperature range and large temperature drops. The surface of some cast concrete was exposed to air for a long time, and was thus affected by the cold. As a result, fractures were likely to be generated early in the life of the concrete.

• In terms of the layout of the site, drainage galleries were arranged in the middle part of the dam where thermal stress is greatest. Accordingly, local stress concentrations caused the top arch to be weak. Meanwhile, the excavation sequence varied, which restricted shrinkage of the concrete and thus led to the concentration of local stresses.

• The concrete in the dam was poorly constructed. Some was uncompacted and contained excess moisture, and some was severely carbonised with the frost-resistant grade failing to reach design requirements. Furthermore, the concrete on the dam faces was damaged by the freeze–thaw process, causing many longitudinally and horizontally distributed penetrating cracks in the body and sluice piers of the dam. In addition, the horizontal construction joints in the dam body were cracked, causing widespread water seepage.

3.4.4.2 Quantitative analysis of the cause of the cracks

Using the open-platform, large-scale commercial software MSC.Marc (www.mscsoftware.com/product/marc), a secondary development of the analytical procedure used for simulating the temperature and stress fields in mass concrete was carried

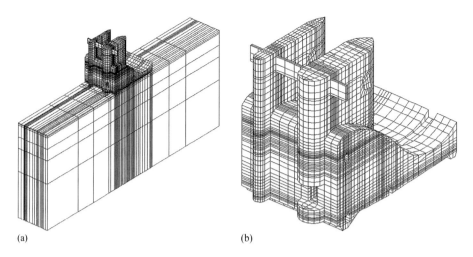

(a)   (b)

Figure 3.24 Finite-element models of the overflow dam (section #17): (a) overflow dam section and bedrock and (b) overflow dam section.

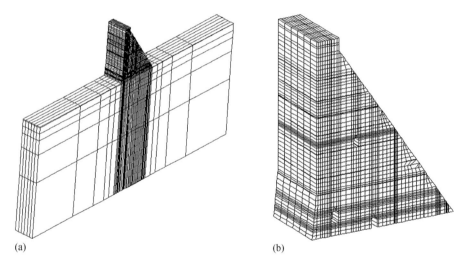

(a)   (b)

Figure 3.25 Finite-element models of the non-overflow dam (section #23): (a) non-overflow dam and bedrock and (b) non-overflow dam.

out. Given the characteristics of this dam, three-dimensional finite-element models of three typical sections in the overflow dam (section #17) and the non-overflow dam (section #23) were established, as shown in Figures 3.24 and 3.25.

3.4.4.2.1 Simulation and analysis of the temperature field during dam operation

Owing to the lack of geothermal data, it was assumed that the boundaries around the foundation were adiabatic. The boundary of the computational domain of the dam

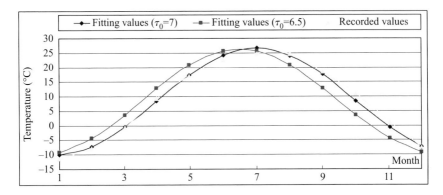

Figure 3.26 Comparison between fitted and measured temperatures.

formed the boundary for any heat exchange. The boundary between the dam and water formed a first-order boundary condition, as demonstrated by the temperature of the reservoir water during normal operation, which is a function of time and position. The boundary between the dam surface and air formed a second-order boundary condition, while that between the surface of the bottom outlets and the atmosphere formed a third-order boundary condition.

Ambient temperature. Because there are no statistical data regarding the temperature in the dam over the years, the actual measured temperatures in the dam from 2005 to 2010 were adopted. The monthly average temperatures in this time period are listed in Table 3.1. By fitting the monthly average temperatures from 2005 to 2010, the following expression for temperature can be obtained (*Specification for load design of hydraulic structures*, SL 744-2016):

$$T_a(t) = T_{am} + A_a \cos\left[\frac{\pi}{6}(\tau - \tau_0)\right] = 8.47 + 18\cos\left[\frac{2\pi}{12}(\tau - 7)\right]$$

where τ is time variable and τ_0 is the initial phase.

The fitted and measured temperatures are compared in Figure 3.26.

Temperature of the upstream reservoir water. There are no measured temperature data for the reservoir water over the years, so the method proposed in the *Temperature Stress and Control of the Mass Concretes* was used to estimate the temperature of the reservoir water.

Calculation conditions

The temperature field was evaluated under three conditions:

- Condition 11: the thin-pier model at normal pool level
- Condition 12: the thick-pier model at normal pool level
- Condition 13: the model of the non-overflow dam sections at normal pool level

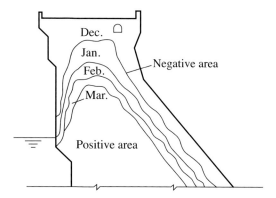

Figure 3.27 Negative temperature regions in the top of the dam sections and the dam body.

3.4.4.2.2 Calculation of the temperature field

The transient temperature fields in some typical months during the operation period of the dam were calculated.

In non-overflow dam section #23, the temperature of the concrete in the dam was found to increase with ambient temperature, and vice versa. The variable temperature on the dam surface remained close to ambient temperature. Inside the dam, with increasing distance from the dam surface, ambient temperature had an increasingly reduced influence on the temperature. For example, the maximum temperatures on the dam surface and at 2 m, 4 m and 6 m from the dam surface were 25.4°C, 18°C, 13.8°C and 11.3°C, respectively, while the minimum temperatures were −8.5°C, −1.46°C, 2.44°C and 4.67°C, respectively. It can be seen that the variable amplitudes of the temperature are 33.9°C, 19.46°C, 11.36°C and 6.63°C. The distributions of the positive and negative temperature regions of each month in the dam are shown in Figure 3.27. It can be seen that there were negative temperature regions inside the dam, away from the surface, from December of each year to March of the next year.

The variation in temperature in thin-pier dam sections is basically the same as that in the non-overflow dam sections. Because the sluice pier is only 4 m thick, the temperature there underwent a synchronous change, with the temperature of its midpoints being slightly lower than that of the surface points of the section. Meanwhile, occurrences of the maximum and minimum temperatures were also slightly delayed. From calculations it was found that the temperature in the thin-pier sections is below zero in January and February.

Although the sluice pier in the thick-pier dam section is 9 m thick, the many pores in it make the concrete in the sluice piers significantly influenced by ambient temperature. The temperature of the section between the working gate and bulkhead gate of the bottom outlets essentially exhibits a synchronous change through its cross-section, and is at negative temperatures in January and February.

3.4.4.2.3 Cause of dam cracks generated during dam operation

Calculation of the load was carried out using the following factors.

- *Water pressure.* The normal pool level water pressure is at 96.6 m, and the design and check flood levels are 101.95 m and 102.6 m, respectively.
- *Uplift pressure.* An upward fold along the surface of the dam foundation was adopted to analyse uplift pressure. The normal stress perpendicular to the surface of the dam foundation calculated using this method was close to the uplift pressure on the surface of the dam foundation. In accordance with the *Safety Assessment Report on the Seepage of Shenwo Reservoir Dam in Liaoning Province*, the strength coefficient of the uplift pressure in the dam sections, including #23 and #17, was taken to be 0.3. The uplift pressure in dam section #25 was not discounted.
- Water pressure exerted by the corresponding water level on the radial gate was imposed on the corbel of the sluice piers in the form of an equivalent concentrated force.
- The water pressure imposed by the corresponding water level on the gate of the bottom outlets was applied to a position near the gate of the bottom outlets in the form of an equivalent concentrated force.
- The load caused by the dead weight of the dam body.
- *Temperature load.* This was calculated by simulating the temperature field of the dam during the operation period. Due to the absence of any data regarding concrete creep in the dam, only the elastic temperature stress was calculated, and the influence of concrete creep was not considered.

The cause of cracking was analysed under the following four conditions:

Condition I: combined model of the thin and thick piers, at normal pool level with consideration of the dead weight of the dam;

Condition II: thin-pier model, at the normal pool level with consideration of the dead weight of the dam and temperature load;

- Condition III: thick-pier model, at the normal pool level and taking the dead weight of the dam and temperature load into account
- Condition IV: model of the non-overflow dam section, at the normal pool level with both the dead weight of the dam and temperature load considered

Simulation was performed to assess the stress field in the dam during operation, as follows:

- At the normal pool level, and under the influence of the dead weight of the dam, except for the stress concentration generated in the corbel of the sluice piers, the combined model of the thin and thick piers showed little stress elsewhere. According to the calculated results, the maximum tensile stress in the corbel of the sluice piers induced by the normal pool level and the dead weight was 1.002 MPa, while the stress in the cracks was only 0.1 MPa. Therefore, it can be seen that the stress induced by the normal pool level and the dead weight was insufficient to produce significant fracturing.
- As discovered in the simulation analysis of the stress field in the non-overflow dam (section #23), because the dam experienced severe cold, the temperature of the dam surface facing downstream was −8.5°C. This led to the generation

of significant tensile stress on the dam surface. The calculated results demonstrated that the tensile stress on the dam surface reached 2.27 MPa, and the tensile stress zone was approximately 3 m thick. In low-temperature periods, the arrival of a cold spell is likely to lead to the generation of cracking.

- Because the thin sluice pier is only 4 m thick, the temperature within its cross-section changed synchronously with ambient temperature. As a result, significant tensile stress was generated in the sluice piers. According to the calculated results, the maximum tensile stress on the surface of the sluice piers was 2.56 MPa, and the tensile stress generated in the middle part of the horizontal section of the sluice piers was 1.17 MPa. That is to say, the whole horizontal section of the sluice piers with an elevation of 85.5 m was stretched, with the tensile stress being more than 1 MPa in cold seasons. If cold spells arrived, this tensile stress would be expected to increase further. It is worth noting was that this was the site where the most, and biggest, cracks were generated.

- Under the influence of ambient temperature, significant tensile stress is expected to be generated in the thick sluice piers due to the existence of many pores therein. It can be seen from the temperature stress calculated from the monthly average temperature that the maximum tensile stress was found both upstream and downstream of the gate slot of the bottom outlets, with their thicker concrete sections, rather than in the gate slot of the bottom outlets. The maximum tensile stress reached 4.5 MPa, which was expected to increase further under the influence of cold weather. Furthermore, when affected by cold spells, significant tensile stress was expected to be generated in the gate slot of the bottom outlets due to the effects of stress concentration.

Above all, cracks generated in the Shenwo dam were mainly caused by thermal stress. Following the generation of these cracks, water could penetrate into the cracks and freeze in winter. As a result, this freezing process led to unstable propagation of cracks.

3.5 Conclusions

Extreme environmental events occur with relatively low frequency. However, with the deteriorating Earth environment and in extreme and abnormal environments, an increasing number of damage events on hydraulic engineering projects have come to light. By taking specific hydraulic projects as examples, the safety operation status of structures under different extreme environmental conditions (earthquake, extreme temperature change and extreme cold environment) has been analysed, and the following conclusions can be drawn:

- By investigating earthquake damage at Shapai and Tongkou hydropower stations after Wenchuan earthquake (magnitude of 8.0), the effects of earthquakes on concrete dams were analysed. The results show the following. 1) Seismic damage to large concrete structures (dam body), abutments, engineering slopes and underground structures is rare and slight, and they have strong earthquake-resistant ability. 2) Secondary disasters following earthquakes have a great impact on buildings, and severe damage as a result of mountain landslides, damming and flooding has been observed. Special attention should be paid to the stability of the natural slope and mountains. 3) Auxiliary buildings, which consist of beams, columns and walls, suffer relatively serious earthquake damage. Specific engineering measures should thus be taken to ensure the engineering safety of subsidiary structures in the earthquake zone.

- By analysing monitoring data from Xixi reservoir during extreme temperature decreases, it can be determined, qualitatively, that low temperatures and high reservoir water levels will lead to a sharp increase in uplift pressure from the dam foundation. By establishing a statistical regression model of the foundation uplift pressure, a quantitative relationship between the uplift pressure and environmental temperature and reservoir water level is detected. The results show that climate fluctuation has an adverse effect on safe operation of concrete dams. To respond to climate fluctuations, some safety measures should be taken to reduce dam operation risk if a sudden temperature drop occurs, such as appropriate control of reservoir water levels and intensifying the monitoring and security checks of the dam.

- Based on analysis of the safe operation of Shenwo reservoir in an extremely cold environment, the following was found. 1) The dam crest tends to shift upstream (horizontal displacement) and seems to lift up year by year (vertical displacement). This phenomenon is mainly induced by freeze–thaw of the concrete in cold environments and corresponding unrecoverable residual deformation leading to abnormal dam displacement. 2) Due to its low frost and crack resistance, the dam has been severely corroded, and the number of cracks in the dam body has been increasing gradually with time. The adoption of the finite element method (FEM) verified that the cracks in Shenwo dam were caused by thermal stress, and frost heave further promoted propagation of the cracks.

References

Bofang Zhu, (2014) *Temperature Stress and Control of the Mass Concretes.* Beijing, China: Tsinghua University Press. (in Chinese)

Cheng-dong Liu, Yan Xiang, *et al.* (2012) *Safety Assessment Report on the Seepage of Shenwo Reservoir Dam in Liaoning Province.* Nanjing, China: Nanjing Hydraulic Research Institute. (in Chinese)

Fuheng Ma, Yan Xiang and Chengdong Liu (2010) *Study on the vulnerability and restoring force of the dam under extreme events.* Nanjing, China: Nanjing Hydraulic Research Institute. (in Chinese)

Houqun Chen, Lincai Dang, Deyu Li, *et al.* (2015) *Code for seismic design of hydraulic structures of hydropower project.* NB 35047-2015: 44–48. (in Chinese)

Houqun Chen, Shunzai Hou, Xirong Guo, *et al.* (1997) *Specification for seismic design of hydraulic structures.* SL 203-97: 28–30. (in Chinese)

Houqun Chen (2009) Consideration on seismic safety of dams in China after the Wenchuan Earthquake[J]. *Engineering Sciences*, 11(6):44–53. (in Chinese)

Jialin Su, Zhigang Hu, Runwei Li, *et al.* (2016) *Specification for load design of hydraulic structures.* SL 744-2016:55–56. (in Chinese)

Jianjun Chi and Xiuli Zhang (2008) Introduction on risk elimination and emergency treatment on hydropower dams in Wenchuan earthquake area. Large Dam & Safety, (3):1–2. (in Chinese)

Jianyun, Z. and Guoqing, W. (2008) Climate change and dam safety. *China Water Resources*, (20): 17–19. (in Chinese)

Jinbao Sheng, Shijun Wang, *et al.* (2012) *Research Report on safety assessment and emergency disposal technology of earthquake damaged reservoir.* Nanjing, China: Nanjing Hydraulic Research Institute. (in Chinese)

Lee, P.S. and You, G.J.-Y. (2011) The risk analysis of long term impact to reservoir under extreme hydrologic events – Shihmen Reservoir, a case study. *Proceedings of World Environmental and Water Resources Congress 2011*: Bearing

Chapter 4
Extreme response of reinforced concrete framed buildings using static and dynamic procedures for progressive collapse analysis

E. Brunesi, G. Fagà and D. Cicola

Recent events have shown that buildings designed in accordance with traditional codes are not necessarily able to resist accidental or man-made extreme events such as impact or explosions. In the past, safety against disproportionate collapse of key structural elements has been increased by means of non-structural protective measures such as barriers, sacrificial components and limitation or control of public access. Codified procedures emerged in the last decades asking for resistant structural design methodologies to inhibit failure incidents acting on the performance of structural components.

In light of this scenario, this chapter presents and discusses an open-access procedure using a fibre-based model in order to reproduce the progressive collapse mechanism of reinforced concrete (RC) buildings subjected to blast loading in an urban environment that leads to the loss of one or more bearing elements. Member removal in this fashion represents an event that happens when extreme situations or abnormal loads destroy the member itself. Two- and three-dimensional models of framed structures have been created and compared using three different numerical tools: an open-source program such as OpenSees and two different commercial codes, SeismoStruct and Ls-DYNA. The first two are more classical fibre-based software, while the last is a well-established general purpose finite element (FE) package. Removal of critical elements is assumed to occur in the building studied and a special-purpose routine has been developed, within OpenSees and SeismoStruct, aiming to create a fibre model capable of simulating the overall structural response due to their failure. In this computational routine, one or more vertical members are instantaneously removed from the model and the ability of the building to successfully absorb member loss is investigated. The results obtained are then compared and validated by using the transient dynamic FE program Ls-DYNA.

The numerical and modelling outcome of this research on progressive collapse behaviour of RC buildings may be immediately applied to the design, vulnerability assessment and strengthening of different structural typologies ranging from residential frames to military facilities.

4.1 Introduction

While numerical modelling is the easiest and most convenient way to represent different aspects of reality using different levels of sophistication, partial or total disproportionate (or progressive) collapse under abnormal loading conditions (e.g. deliberate terrorist attacks, uncontrolled gas releases or vehicle or aircraft impacts) is one of the most vivid examples of a low-probability–high-consequence event that may occur in the lifetime of a structure. Indeed, several iconic and public buildings that have been subject to accidental and man-made extreme events in recent decades have resulted in significant casualties and property loss due to the fact that structural systems designed according to conventional approaches are not necessarily able to withstand them (Brunesi and Nascimbene, 2014; Brunesi et al., 2015). In light of this, homeland security has become a primary concern for public authorities and stakeholders, seeking for passive and active strengthening strategies of different structural typologies ranging from civil to strategic and military facilities.

A few studies have been carried out emphasising the need for probabilistic risk assessment and management of structures in relation to disproportionate collapse (Ellingwood, 2006; Park and Kim, 2010; Asprone et al., 2010; Brunesi et al., 2015) and several definitions of this phenomenon have appeared in the literature (Ellingwood and Leyendecker, 1978; Gross and McGuire, 1983; Corley et al., 1998; Nair, 2006; Bažant and Verdure, 2007; Starossek, 2007; Starossek and Haberland, 2010), while blast- or progressive collapse-resistant building analysis and design have been recognised to have an increasing impact on economy and society.

To ensure resistance against disproportionate collapse phenomena, a structure should respect five main requirements (Wibowo and Lau, 2009): robustness, integrity, continuity, redundancy and ductility. In accordance with the classifications and definitions given by Starossek (2009), robustness is purely a property of the structure, in contrast with a more general perspective provided in Eurocode UNI EN 1991-1-7 (CEN, 2006), which refers to a broader triggering of accidental actions (i.e. the Eurocodes do not include a separate 'structural' standard for progressive collapse, but they include it in the Accidental Action code). Integrity refers to the ability of the structural connections between members to carry loads after the occurrence of abnormal events. The document published in February 2007 by the United States National Institute of Standards and Technology (NIST, 2007) offers an overview of different approaches to structural integrity, and Menchel (2009) proposed a review of available standards for design against progressive collapse, such as those released by the General Services Administration (GSA, 2003) and the Department of Defense (DoD, 2005). Continuity defines the interconnectivity between structural elements such as beams, columns and slabs. ASCE 7-02 (2002) requires that the structural integrity be achieved by providing sufficient continuity, redundancy and ductility in the members of the structure. The existence in a building of alternative load paths for force transfer is usually referred to as redundancy; this simply implies the capability of 'other' structural members, different from the one collapsed, to carry extra load and to balance the extreme demand induced by sudden element loss scenarios. Finally, the term ductility refers to the ability of a structural system (i.e. element, section or material) to deform beyond elastic limits without excessive stiffness degradation or strength deterioration (Paulay and Priestley, 1992).

4.2 Direct and indirect design methods

When designing (or verifying) a structure to be less vulnerable to progressive collapse, the the five main requirements listed above must be comprehensively considered by following two main methods: direct and indirect. The direct design method explicitly provides resistance to the structure by enhancing the strength of key elements (i.e. by preventing a local failure assuming a specific local resistance; NIST, 2007; Starossek, 2009) or by designing the skeletal frame to bridge across collapse (i.e. by assuming a local failure using alternative load paths; GSA, 2003; DoD, 2005). While the direct approach relies explicitly on structural analysis and design (Saad *et al.*, 2008), the indirect method considers resistance to progressive collapse implicitly through prescriptive design rules, which are intended to provide minimum levels of ductility and continuity (National Research Council of Canada, 1996; British Standards Institution, 1997; ASCE 7-02, 2002; DoD, 2005).

In accordance with the review carried out by Dusenberry and Juneja (2003) and the description discussed by Starossek (2009), the following common prescriptive rules must be satisfied in the building design:

- As specified by European (CEN, 2006) and US (ACI 318, 2002) codes, both horizontal and vertical tie elements (such as ordinary steel cables or post-tensioned strands) should be provided to transfer tensile forces and enhance overall integrity (Li *et al.*, 2011).
- In the case of intermediate column failure, a transition from flexural to tensile load transfer happens. Beam-catenary (or slab-membrane) action should be enabled in order to activate a bridge over the failed column and thus to provide continuity within structural members (Valipour and Foster, 2010; Dat and Hai, 2013). In RC sections this could be done by using composite sections or more classical seismic details such as the continuity of top/bottom reinforcements over a failing column.
- When a major abnormal load imposes large deformation, the structure should be capable of sustaining a high fraction of the initial strength. This ability of the building or its elements or its sections or its materials to be beyond the elastic limit is usually referred to as ductility (Paulay and Priestley, 1992).

4.2.1 Numerical analysis approaches

In both direct and indirect procedures, four analyses can be used according to the classification described by Marjanishvili (2004):

- Linear static analysis
- Nonlinear static analysis
- Linear dynamic analysis
- Nonlinear dynamic analysis

The disadvantage of linear analyses, both static and dynamic, is the inability to include material and geometric nonlinearities such as large displacements/rotations (i.e. beam-catenary action), second-order effects, inelastic behaviour and plastic hinge formation (i.e. strength or stiffness degradation and ductility). Nonlinear static (or pushdown) analysis is relatively simple and gives a capacity curve that, similar to a seismic analysis, provides insight into whether a building has adequate capacity to resist an extreme loading condition or not, in a static fashion. One determining factor in considering that a local portion of the structure has failed is the highly dynamic effect produced when a structural element is rapidly removed from the frame. As demonstrated by Pretlove *et al.* (1991), there are structures that are statically safe, but dynamically unsafe, due to the fact that the time-dependent overloads induced by element removal may cause the progressive fracture of other elements before a new equilibrium state is reached (i.e. cascade or domino effects). This requires the nonlinear dynamic behaviour of a structure to be taken into account in progressive collapse simulations.

Considering the aforementioned observations regarding the five main structural requirements (i.e. robustness, integrity, continuity, redundancy and ductility), two design methods (i.e. direct and indirect approaches) and two sources of inelasticity (i.e. geometric and material nonlinearities), the flowchart presented in Figure 4.1 was constructed to describe the numerical open tool developed within OpenSees (2013) and SeismoStruct (2013).

A wide range of nonlinear fibre-based implicit dynamic simulations were carried out to examine the responses of two- and three-dimensional RC framed structures, and the analysis results were then compared with a Hughes–Liu (1981) FE Ls-DYNA (2010) model, which was analysed in explicit dynamic fashion to prove the ability of the proposed approach. In light of this, a further aim of this chapter is the development of a computational methodology for structural design against progressive collapse to be implemented into an open-source platform.

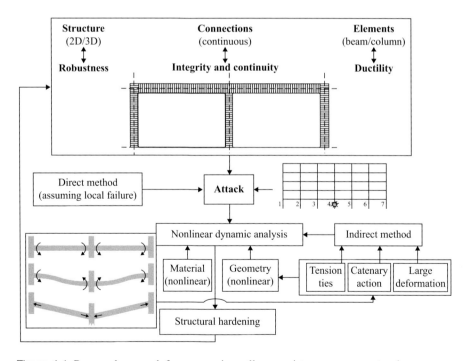

Figure 4.1 Proposed approach for progressive collapse resistance assessment and strengthening strategy.

4.3 Numerical models for progressive collapse assessment

To withstand abnormal loading that may cause progressive collapse, there are several characteristics to be fulfilled in progressive collapse simulations. Many commercial software packages can be used for this purpose and some have specific options for progressive collapse simulations. In the past two decades, several modelling applications have been developed:

- Three-dimensional models using four-node quadrilateral shell elements in ABAQUS (Lee *et al.*, 2009; Wei *et al.*, 2007)
- Macromodels based on nonlinear springs to obtain the nonlinear static response of single beams following column removal, using the code ADAPTIC (Vlassis *et al.*, 2008)
- Two-dimensional boundary-element models coupled with two-dimensional FE models using ADINA (Kripakov *et al.*, 1995)
- Three-dimensional solid elements coupled with three-dimensional Euler flux-corrected transport (FCT) for the air volume using AUTODYN (Luccioni *et al.*, 2004)
- 20-node brick elements with a total Lagrangian formulation to model beam-column subassemblies in DIANA (Bao *et al.*, 2008)
- A finite difference approach to reproduce structural concrete and steel connections using DYNA3D (Krauthammer, 1999)
- Multibody models based on bricks and continuum-based multilayer shell elements in FEAP and Ls-DYNA (Möller *et al.*, 2008; Hartmann *et al.*, 2008)
- Solid and shell elements to simulate a three-storey two-span RC frame with initial damage to structural members using Ls-DYNA (Shi *et al.*, 2010)
- Macromodels based on rigid elements and inelastic shear links to model concentrically and eccentrically steel-braced frames using Ls-DYNA (Khandelwal *et al.*, 2009)
- Bernoulli beam elements with concentrated plastic hinges aprioristically assigned to all possible locations where stiffness and strength degradation can occur, using SAP2000 (Sasani, 2008; Tsai and Lin, 2008)

Beam FEs have the advantage of reducing computational costs, but most past attempts have been confined to two-dimensional subsystems (Usmani *et al.*, 2003; Kaewkulchai and Williamson, 2004; Kim *et al.*, 2009) and numerous simplifications are applied, such as the trilinear model for RC compact sections (Isobe and Tsuda, 2003) or lumped plasticity assumed to occur only at the element ends (Kaewkulchai and Williamson, 2004; Talaat and Mosalam, 2008; Menchel, 2009). By contrast, detailed brick elements with embedded steel reinforcements and shell elements (Shi *et al.*, 2010) may give more insight, but tremendously increase the required computational time, needing parallel processing on multiprocessor computers.

A third way to model is through a fibre-based force (Spacone *et al.*, 1996) or displacement (Hellesland and Scordelis, 1981) approach, as is commonly used in the seismic analysis of steel structures (Wijesundara *et al.*, 2009, 2011), RC buildings

(Mpampatsikos *et al.*, 2008), masonry panels (Smyrou *et al.*, 2011) or connections (Nascimbene *et al.*, 2012; Brunesi *et al.*, 2014) and continuous span bridges (Casarotti and Pinho, 2006). A first application of fibre beam models based on the general-purpose commercial FE package MSC.MARC has been carried out by Lu *et al.* (2008) and then used to improve the tie force method adopted by BS 8110-1:1997 (1997), Eurocode 1 (CEN, 2006) and the DoD (2005), including load redistribution in three dimensions, dynamic effects and internal force correction (Li *et al.*, 2011).

Recently, Kim *et al.* (2009) applied the open-source code OpenSees for the development of an integrated system of progressive collapse analysis. Two- and three-storey two-dimensional framed structures were analysed, but assumed that concentrated plastic hinges form only at the ends of structural elements. Therefore, one of the main objectives of the presented study is to develop an appropriate procedure for large-scale nonlinear transient dynamic analysis of three-dimensional frames, combining the force-based fibre element with the open-platform OpenSees. The automatic element removal algorithm is formulated in light of an object-oriented architecture and the main capabilities of the computational solution strategy proposed herein can be categorised into two levels: local/element and global/structural.

4.4 Reference framed structure

In the following, the prevailing geometric characteristics and assumptions related to the analysed structural configuration will be discussed, as well as its FE representation and the column-removal strategy used to assess the two- and three-dimensional idealisation of the reference framed structure under investigation.

4.4.1 Building description and structural assumptions

The prototype framed building used in the study was a five-storey 6×6-bay RC building. The floor span and height were 6 m and 3 m, respectively. As shown in Figure 4.2, the out-of-plane span was assumed to be 4 m.

Column and beam cross-sections were 400×400 mm² and 500×300 mm², respectively. A uniform longitudinal reinforcement (8 $\emptyset$18) was provided in the columns at each floor, while 7 $\emptyset$18 at the top and 5 $\emptyset$18 at the bottom of the cross-section were used to reinforce all the beams ($\emptyset$18 stands for 18 mm diameter rebars). The transverse reinforcement was composed of $\emptyset$8 stirrups, spaced at 20 cm, in either the beams or columns. This arrangement represents an appromately 80% increase with respect to that demanded by a conventional static design, conducted by assuming dead and live loads of 6 kN/m² and 3 kN/m², as the building was assumed to be used for offices. The downward load (Q_b) given in equation 4.1 was used for progressive collapse assessment both in static and dynamic modes:

$$Q_b = DL + 0.25LL = 27 \text{ kN/m} \qquad (4.1)$$

where *DL* and *LL* stand for dead load and live load, respectively.

Although a properly detailed slab may potentially improve the structural response by participating in the redistribution mechanism of the entire building, the one-way RC floor slab was conservatively assumed to provide no resistance against progressive collapse, but its weight and inertia were implicitly included in the FE simulations

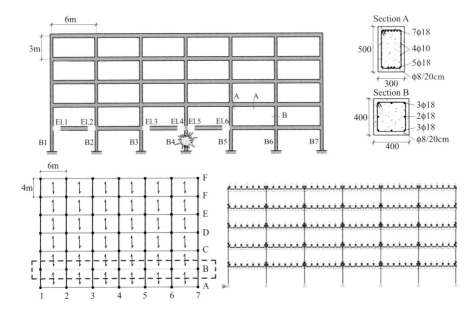

Figure 4.2 Reference MRF: geometry, elements and frames nomenclature, cross-sections and reinforcement arrangements, plan view and two-dimensional FE representation.

Table 4.1	Element acronym and position.			
Acronym	Frame	Floor	Beam	Position (m)
El.1			B1-B2	0.2
El.2				5.8
El.3		1st	B3-B4	12.2
El.4	B			17.8
El.5			B4-B5	18.2
El.6				23.8
El.7		2nd	B3-B4	17.8

carried out. The material properties were assigned with 25 MPa for the concrete compressive strength and 450 MPa for the yield strength for both the longitudinal and transverse reinforcement. Table 4.1 identifies the key elements, their position and the acronym used in the upcoming discussion of the numerical results obtained.

4.4.2 FE representation of the structure

Inelastic force-based fibre elements, with five integration points, were used within Open-Sees and SeismoStruct, while the approach proposed by Hughes and Liu (1981), which is based on a degenerated brick element formulation including shear strains, was used in Ls-DYNA. Even though the use of a force-based formulation does not necessarily imply the need for element discretisation, a one-to-six correspondence between structural members and model elements was assumed in the analysis to accurately accommodate the deformed shape; mesh sensitivity was then checked by halving this ratio.

Geometric nonlinearity was taken into account using a co-rotational transformation, the implementation of which is based on an exact description of the kinematic transformations associated with large displacements and three-dimensional rotations of the beam-column member. A similar technique was implemented in the three FE packages considered in this research to allow for the large displacements/rotations expected during progressive collapse simulation.

Material nonlinearity was described in each program by a distributed inelasticity approach, in which the sectional stress–strain state of each beam–column element is obtained through the integration of the uniaxial stress–strain response of the individual fibres. In detail, each cross-section was discretised in 400 fibres and equivalent stress–strain relationships were selected within the three numerical tools. In Seismo-Struct, the classical uniaxial uniform confinement model proposed by Mander et $al.$ (1988) was used to represent the inelastic behaviour of concrete, additionally accounting for tension softening. A bilinear idealisation, combined with isotropic strain hardening, was assumed for steel reinforcement. Equivalent constitutive laws are available in OpenSees and, hence, advantage was taken of Concrete07 and Steel01 materials. Finally, MAT_174_RC_BEAM was adopted in Ls-DYNA to equivalently reproduce the aforementioned constitutive laws. According to Bischoff and Perry (1991), strain rate effects were conservatively neglected in the simulation.

Death element option was used to operate the member removal in Ls-DYNA, while a special-purpose routine was prepared and integrated in OpenSees and SeismoStruct to simulate the sudden column loss. In particular, the numerical technique proposed collects the actions at the node where the column is supposed to be lost, statically applies them as reactions after element removal, and dynamically cancels them by means of forces with the same magnitude but opposite sign.

A displacement/rotation-based convergence criterion, with a threshold set equal to 10^{-3}, was adopted to equilibrate loads through an incremental iterative procedure; both Newton–Raphson and Krylov subspace algorithms were tried, because the latter has a larger radius of convergence and requires fewer matrix factorisations (Scott and Fenves, 2010). In SeismoStruct and OpenSees, an integration time step of 2×10^{-4} was used to perform a series of implicit dynamic simulations, while an explicit solution strategy, with an automatic mesh-dependent integration time step on the order of 10^{-5}, was adopted in Ls-DYNA. In accordance with Priestley and Grant (2005), the tangent stiffness-proportional Rayleigh damping model was assumed to conduct the nonlinear dynamic analyses using each of the three FE codes.

Conventional performance criteria were then defined in terms of strain limits for concrete and steel; in particular, a value of 5×10^{-3} was assumed as the ultimate concrete compressive strain, accounting for the confinement effect provided by the transverse reinforcement (Mander et $al.$, 1988), and a value of 6×10^{-2} was conservatively assumed as the ultimate steel strain. Hence, the failure of the RC framed structure was supposed to occur when one of two conditions was met:

- First exceedance of the first strain limit in the first section of the frame
- Divergence of the solution procedure or displacement–time history curve.

In the series of progressive collapse simulations discussed in the following, numerical and structural failure criteria were observed to coincide, because exceeding these strain limits at the sectional level implied large plastic rotational demands at the element level, which in turn caused the building to collapse, as it was too damaged to oscillate around a peak displacement.

4.4.3 Loading procedure and assessment strategy

First, an internal two-dimensional frame, referred to as model B, was extracted from the entire building and its response was assessed by the three FE programs to achieve independence of the results with respect to analysis type and algorithm, as well as to element formulation and removal technique. A comparison was then made with the three-dimensional representation of the case-study prototype to investigate the capabilities of the secondary frames system, which were recognised to be crucial in the creation of a rationally controlled alternative load path for the unbalanced demand caused by the sudden column loss (Brunesi and Nascimbene, 2014; Brunesi *et al.*, 2015). This was essentially intended to make them work in the structural scheme as a latent resource for stiffening and strengthening, thus improving the progressive collapse potential of the building.

The selected six threat-independent column-removal conditions, named in the following as scenarios B1, B2, B3, B4, B2-4-6 and B3-4 according to bay line letters and numbers (Figure 4.2), consider the failure of each single column, as well as the simultaneous loss of both three alternate and two consecutive columns. This loading strategy was then extended to the three-dimensional case, for a total of 54 configurations, which were identified by removing columns belonging to more than one frame at once. Given the double symmetry in plan, the analyses were focused on investigating only a quarter of the building under study.

4.5 Nonlinear FE simulations: results and discussion

The numerical results obtained in accordance with the aforementioned strategy will be discussed in the following sections, in order to show and quantify the overestimates of two-dimensional models for progressive collapse resistance assessment of three-dimensional skeletal frames. This permitted correlating the structural redundancy added by the secondary beam system and the displacement reduction observed in the case of a three-dimensional simulation.

4.5.1 Two-dimensional models

A series of nonlinear dynamic analyses for the column removals (specified above) were carried out, accounting for large displacements/rotations via co-rotational geometric transformation, and both vertical displacement and velocity–time histories were collected (Figure 4.3). Good agreement between the predictions obtained by the three codes is achieved, with discrepancies lower than 1% in terms of peak displacement. The most demanding condition is B1, which results in a maximum deflection of about 17 cm, while the other cases range from about 10 to 11 cm. Similar levels were shown by Santafé Iribarren *et al.* (2011) for similar two-dimensional RC frames. Velocity peaks of up to 750 mm/s have to be accommodated. Even though Ls-DYNA and SeismoStruct show a slight mismatch with the OpenSees response

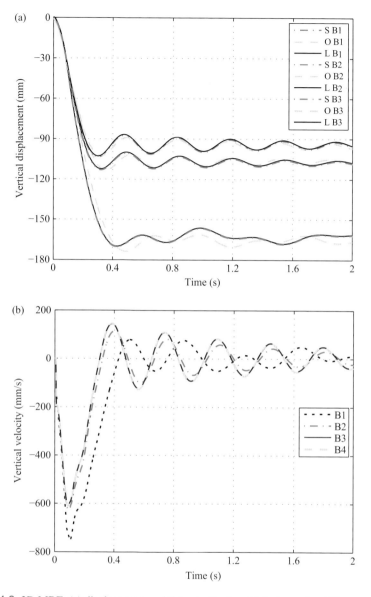

Figure 4.3 2D MRF: (a) displacement and (b) velocity time histories under different column removal conditions. S, SeismoStruct; O, OpenSees; L, Ls-Dyna.

due to a slightly different implementation of the Mander model in the post-peak and unloading branches, particularly at large strains, a quite similar trend can be observed for B2, B3 and B4. In particular, B3-4, B-2-4-6 and B4 are not reported in Figure 4.3a, because the analysed moment resisting frames (MRFs) collapsed in the case of both two consecutive and three alternate columns, while B4 and B3 revealed an almost identical response, as confirmed by Figure 4.3b.

An example of the extreme progressive collapse-induced demand is shown in Figure 4.4a, which presents the time-dependent evolution of the bending moment distribution observed in the beams of the first floor when the central column is removed. In particular, moment demand is collected after column removal, at peak displacement, at the end of the analysis and for two intermediate time instants (pre- and post-peak displacement). According to the almost symmetrical reinforcement provided in the beams, a maximum bending moment, slightly lower than 400 kNm, was

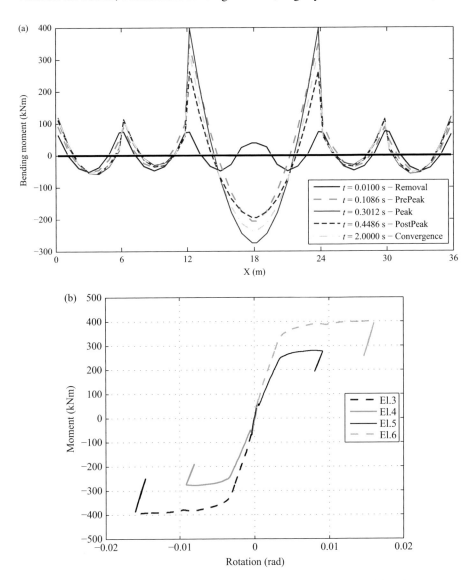

Figure 4.4 Central ground column removal: (a) time-dependent evolution of bending moments in the beams of the first floor; (b) corresponding moment–rotation and (c) axial force–vertical displacement curves in the key elements.

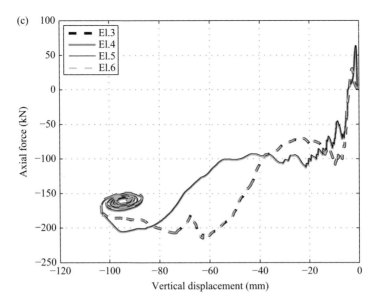

Figure 4.4 (*Continued*)

experienced, in correspondence to the beam end opposite to the column removed, while a peak equal to approximately 75% was determined above B4. In addition, the ratio between bending moment at peak displacement and that observed at column removal is larger than 5.5. The resisting mechanism provided was confirmed by moment–rotation curves in the critical beam sections, as shown in Figure 4.4b, while the arch effect and catenary action, developed in the beams of the first and second floors, are highlighted in Figure 4.4c. In particular, the change in bending moment sign produces a reduction in the demand, as the RC cross-section was characterised by a similar capacity under loading reversals. The moment–rotation curve reaches its 'unloading' branch and oscillates upward and downward when the structure oscillates around its displacement peak.

After column removal, an arch mechanism develops between the top of the middle joint and the bottom of the exterior joint, resulting in axial compressive forces in the beams. Stresses increase until concrete crushing and cracking occur in the compressive and tensile regions, respectively, resulting in a plastic hinge that was observed to limit sectional carrying capacity. Cracking further develops through the beam depth, and catenary action decreases the compressive forces. Moderate tension appears in the beams of the second floor, while those at the first floor are observed to fluctuate, being further damaged and, hence, unable to react.

A series of incremental-mass nonlinear dynamic analyses were thus performed by factorising the load combination given in equation 4.1 in order to assess any further resources against progressive collapse that the structure may potentially exploit. In particular, a step function, whose magnitude was monotonically increased until failure occurred, was used to simulate the different dynamic loading conditions. The displacement–time curves obtained for B1 are presented in Figure 4.5a, while Figure

4.5b shows the corresponding moment distributions at peak displacements for the dynamic load factors imposed. The two-dimensional representation of the structure, analysed in this column removal configuration, collapsed for Q_b greater than unity, as it was unable to accommodate larger plastic rotation demands in 'El.2'. As highlighted by the moment–rotation curves at ultimate conditions, shown in Figure 4.5c, a peak

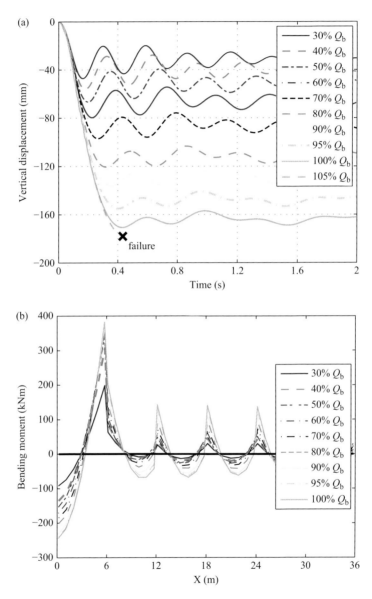

Figure 4.5 Incremental-mass nonlinear dynamic analyses in the case of leftmost column removal: (a) displacement time histories; (b) evolution of bending moments; (c) moment–rotation curves at ultimate conditions.

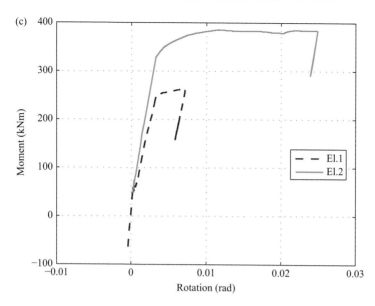

Figure 4.5 (*Continued*)

of about 0.025 rad, which is close to the capacity of the cross-section, is observed. By contrast, the maximum loading capacity was $1.05Q_b$, $1.10Q_b$ and $1.10Q_b$ for cases B2, B3 and B4, respectively, because the progressive collapse-induced mechanism took advantage of two adjacent beams to redistribute the unbalanced loads. Finally, $0.95Q_b$ was reached for B2-4-6, as the overloads above B4 were counterbalanced by those acting on B2 and B6.

The displacement peak of each analysis was then collected to construct load–displacement capacity curves, such as that shown in Figure 4.6a, where the envelope obtained from the series of incremental dynamic analyses is compared to the results of a displacement-controlled nonlinear static (or pushdown) analysis performed by applying uniform loads along the beams (GSA, 2003). Obviously, the nonlinear static analysis overestimates the ultimate capacity by roughly 15%, as confirmed by the dynamic amplification factor (DAF) vs. displacement plots, presented in Figure 4.6b. Hence, the assumption of a constant force-based DAF equal to 2, as prescribed by conventional procedures in current codes (GSA, 2003), was reaffirmed to be over-conservative, particularly if the design target moves into the nonlinear branch of the framed building. Under the displacement demand of collapse resistance, DAF may decrease to 1.11–1.13. Similar results have been obtained for high-rise buildings by Tsai and Lin (2008).

4.5.2 Three-dimensional models

Once the robustness of an internal two-dimensional framed structure had been assessed, the entire building under investigation was analysed by using a campaign of nonlinear dynamic analyses, which were explicitly carried out to highlight the positive influence of the orthogonal frames. This was mostly to show that, if properly designed

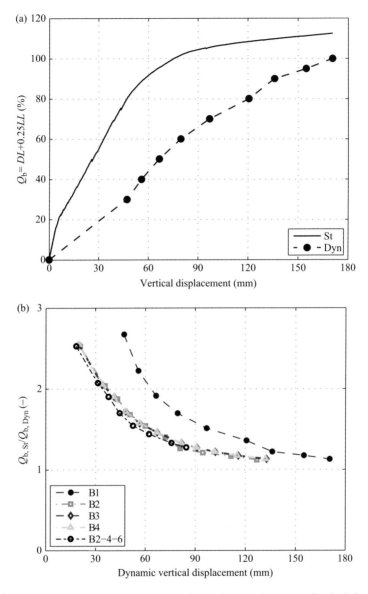

Figure 4.6 Nonlinear static versus dynamic analyses: (a) capacity curves for the leftmost column removal and (b) DAF.

and detailed, the additional redundancy of these framing systems may provide an optimal combination of stiffness, strength and ductility for a traditional three-dimensional RC structure. The orthogonal beams are expected to sustain only their own weight, because the slab is unidirectional and, hence, their design is governed by minimum requirements, ordinarily prescribed in conventional codes. In particular, the cross-section depth was kept constant, and 1% of longitudinal reinforcement, symmetrically arranged, was imposed to satisfy minimum seismic requirements (CEN, 2004), always mandatory in European Standards. This results in a flexural resistance roughly equal to one-half of that provided in the primary beams. Therefore, the redundancy added was observed to control the structural response, providing an alternative load path. Figure 4.7 presents both displacement and velocity histories for some significant column removal conditions, while Table 4.2 (page 93) summarises the peaks obtained in terms of vertical displacement ($D_{max,3D}$) and bending moment demand in the beams of both primary ($M_{max,P}$) and secondary ($M_{max,S}$) frame systems. In addition, their ratio with respect to those observed in the static condition ($R_{M,P}$ and $R_{M,S}$) was computed

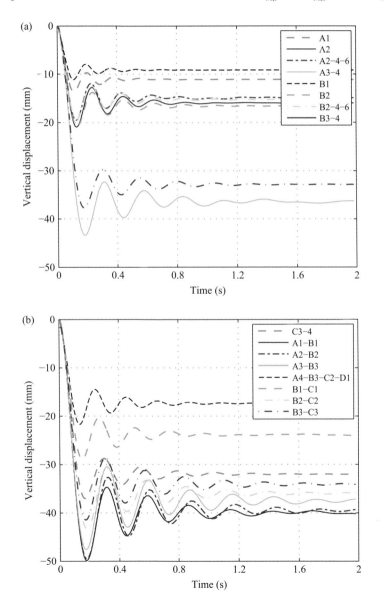

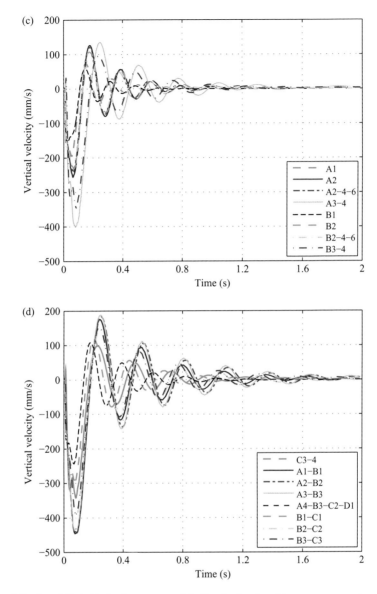

Figure 4.7 3D MRF: (a,b) displacement and (c,d) velocity time history curves under significant ground column removal conditions, according to the loading strategy selected.

and collected. Finally, normalisation was proposed with respect to the peaks determined through the two-dimensional representation of the reference building.

Displacement peaks ranging from 11 to 21 mm are shown in the case of a single column removal, while the related velocities decrease to about 150–250 mm/s. This significant reduction leads the three-dimensional framed building to behave almost elastically. If frame A is taken as reference, a significant decrease (55–65%) is

Table 4.2 Significant quantities obtained from a three-dimensional representation of the case-study prototype.

Case	$D_{max,3D}$	$M_{max,P}$	$M_{max,S}$	$R_{M,P}$	$R_{M,S}$	$D_{max,3D}/$ $D_{max,2D}$	$M_{max,P}/$ $M_{max,2D}$	$M_{max,S}/$ $M_{max,P}$
	(mm)	(kN m)	(kN m)					
A1	13.7	131.5	119.1	2.84	13.30	0.080	0.339	0.906
A2	21.0	175.7	161.2	3.81	18.02	0.123	0.453	0.918
A2-4-6	19.6	185.2	159.3	4.04	17.88	0.115	0.478	0.860
A3-4	43.5	234.2	189.9	5.07	21.17	0.255	0.604	0.811
B1	11.1	172.4	120.0	2.23	13.40	0.065	0.445	0.696
B2	20.8	238.6	169.5	3.09	17.35	0.122	0.616	0.710
B2-4-6	19.5	252.6	168.7	3.28	17.28	0.114	0.652	0.668
B3-4	37.8	272.6	183.4	3.53	18.89	0.221	0.703	0.673
C3-4	37.1	271.3	189.5	3.51	21.17	0.218	0.700	0.698
A1-B1	50.0	322.1	201.7	4.16	22.52	0.293	0.831	0.626
A2-B2	49.6	348.8	203.9	4.51	21.99	0.291	0.900	0.584
A3-B3	47.7	357.1	203.1	4.62	22.07	0.280	0.921	0.569
A4-B3-C2-D1	21.7	245.7	170.8	3.18	17.59	0.127	0.634	0.695
B1-C1	29.0	276.9	177.4	3.58	19.82	0.170	0.714	0.641
B2-C2	43.3	343.4	198.0	4.44	20.39	0.254	0.886	0.577
B3-C3	41.6	351.6	195.9	4.55	20.32	0.244	0.907	0.557

experienced in terms of moment in the primary beams, according to the remarkable redistribution effects provided by the secondary frame elements, characterised by only 10% lower peaks. This reduction is slightly less visible in the internal frames (e.g. frame B). Furthermore, the three-dimensional structural system is able to sustain the case of A2-4-6 by accommodating displacements slightly lower than those observed in the case of A2. Even the simultaneous removal of two consecutive columns of a single frame (e.g. A3-4) was passed, although it proved to be much more demanding. In comparison with A3, more than doubled displacements were predicted, thus implying a larger exploitation of the reserves given by the primary frame A (i.e. $R_{M,P}$ of 5). A similar trend was shown by B3-4 and C3-4, even though 15% lower displacements had to be applied, because two adjacent secondary frames participated in the redistribution mechanism in this case. To further point out and highlight their contribution, case A4-B3-C2-D1 presented a response identical to B3. Finally, a different behaviour is shown if the same column is removed from different parallel frames (e.g. A1-B1). Vertical displacements up to 50 mm, corresponding to almost four times those experienced in the case of A1, were determined. In addition, 60% and 40% increments were observed for $M_{max,P}$ and $M_{max,S}$, respectively, and the largest $R_{M,S}$ was predicted in this case. Similar conclusions can be drawn for the other cases of this removal

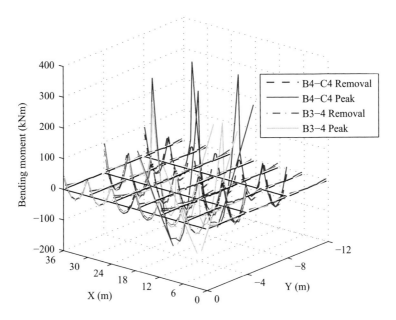

Figure 4.8 B4-C4 versus B3-4: bending moment distributions at column removal and peak vertical displacement.

configuration. However, those involving frames B and C presented lower displacement and moment demand, because the alternative load path took advantage of two orthogonal frames instead of just one.

To emphasise the key role played by the secondary system in providing an additional contribution to the primary structural scheme, cases B4-C4 and B3-4 were compared, as shown in Figure 4.8, thus depicting their moment distribution at column removal and peak displacement. A larger displacement, close to 42 mm, is achieved if the B4-C4 configuration is considered, while a peak of about 38 mm is observed in the case of B3-4, resulting in a 9% reduction. A similar effect is seen by looking at moment demand in the primary and secondary frames (22% and 7% decreases are observed, respectively). Hence, the most demanding of the two conditions was proven to be that involving the simultaneous removal of the same column on two adjacent primary frames rather than that implying the removal of two adjacent columns in the same primary frame, at once. In fact, in the case of B3-4, both columns are connected to two secondary beams, while the B4-C4 configuration exploits, for redistribution purposes, only one out-of-plane beam, thus revealing their potential for developing alternative load paths, if properly designed.

To quantify the reserves of the secondary frame system, a trend may be established between the normalised displacement ($D_{max,3D}/D_{max,2D}$) and the demand-to-capacity ratio of its beams ($M_{max,S}/M_{Rd,S}$), as shown in Figure 4.9 and analytically expressed in equation 4.2:

$$\frac{D_{max,3D}}{D_{max,2D}} = 7.7 \times 10^{-3} e^{3.6812 \times \frac{M_{max,S}}{M_{Rd,S}}} - \quad R^2 = 0.92 \qquad (4.2)$$

94

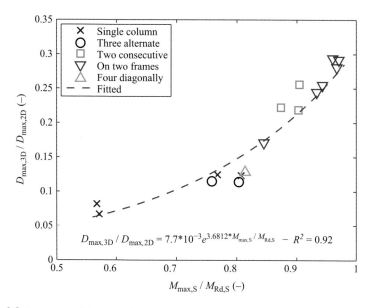

Figure 4.9 Reserves of the secondary frame system: $D_{max,3D}/D_{max,2D}$ versus $M_{max,S}/M_{Rd,S}$.

The exponential expression, determined from a regression analysis using the least-squares technique, gives, for the reference building, a measure of the displacement reduction in the case of three-dimensional analysis as a function of the required target for the secondary beams, which may potentially be fixed at a preliminary design stage. A satisfying fit is achieved in this case, with a coefficient of determination (R^2) equal to 0.92.

4.6 Conclusions

A procedure for simulating the large displacement inelastic dynamic response of RC framed buildings subjected to sudden column loss was implemented into an open-source fibre-based code and then verified by comparison with general-purpose FE software. The independence of the results with respect to analysis type and algorithm, as well as element formulation and removal technique, was achieved by assessing an internal two-dimensional frame extracted from a five-storey 6 × 6-bay RC structure. A progressive collapse resistance assessment was conducted using nonlinear static and dynamic analyses, which were carried out either using implicit or explicit solution procedures. The load–displacement capacity curves determined from the two methods were compared, revealing the former to be conservative by roughly 15% at ultimate conditions. The decay of the force-based DAF throughout the overall inelastic response of the frame was captured, proving current codified procedures to be over-conservative.

Three-dimensional simulations were also conducted to assess the behaviour of the entire skeletal frame investigated, thus establishing the pros and cons of two-dimensional models. In particular, the secondary frames system, with 'ad hoc' design and detailing, was demonstrated to significantly contribute to the structural response, acting as a latent resource for stiffening and strengthening. Therefore, a rationally controlled alternative load path for the unbalanced demand caused by the sudden column loss was feasible at the design stage and may take advantage of the structural redundancy demanded by modern seismic codes in order to ensure the three-dimensional behaviour of a structure. A trend was then established between displacement reduction in the case of the three-dimensional simulation and the demand-to-capacity ratio of the secondary beams, which were specifically designed to be weaker than the primary framing system. Several removal conditions were discussed to determine and quantify the trend observed, using them as a sort of incremental load to push the three-dimensional framed building to the limit of its capacity.

Ultimately, the use of minimum seismic requirements, as prescribed in any of the current codes, was revealed to be a promising technique for a cost-effective strengthening strategy to mitigate the risk associated with disproportionate collapse of new constructions under abnormal loading conditions. Additional redistribution capabilities for progressive collapse could be studied in properly designed and detailed floor slabs or infill systems, whose resistance is expected to provide a significant potential for alternative load paths. In light of this, a structural skeletal frame should benefit from the contribution of further secondary systems.

References

American Concrete Institute (2002) *Building code requirements for structural concrete (ACI 318-02) and commentary (ACI 318R-01)*. American Concrete Institute, Farmington Hills, MI.

American Society of Civil Engineers (2002) *Minimum design loads for buildings and other structures*, ASCE 7-02. ASCE, Reston, VA.

Asprone, D., F. Jalayer, A. Prota and G. Manfredi (2010) Proposal of a probabilistic model for multi-hazard risk assessment of structures in seismic zones subjected to blast for the limit state of collapse. *Structural Safety* **32**, 25–34.

Bao, Y., S.K. Kunnath, S. El-Tawil and H.S. Lew (2008) Macromodel-based simulation of progressive collapse: RC frame structures. *Journal of Structural Engineering* **134**, 1079–1091.

Bažant, Z.P. and M. Verdure (2007) Mechanics of progressive collapse: learning from World Trade Center and building demolitions. *Journal of Engineering Mechanics* **133**, 308–319.

Bischoff, P.H. and S.H. Perry (1991) Compressive behaviour of concrete at high strain rates. *Materials and Structures* **24**, 425–450.

British Standards Institution (1997) *Structural use of concrete. Code of practice for design and construction*, BS 8110-1:1997. BSI, London.

Brunesi, E. and R. Nascimbene (2014) Extreme response of reinforced concrete buildings through fiber force-based finite element analysis. *Engineering Structures* **69**, 206–215.

Brunesi, E., R. Nascimbene and G.A. Rassati (2014) Response of partially-restrained bolted beam-to-column connections under cyclic loads. *Journal of Constructional Steel Research* **97**, 24–38.

Brunesi, E., R. Nascimbene, F. Parisi and N. Augenti (2015) Progressive collapse fragility of reinforced concrete framed structures through incremental dynamic analysis. *Engineering Structures* **104**, 65–79.

Casarotti, C. and R. Pinho (2006) Seismic response of continuous span bridges through fiber-based finite element analysis. *Earthquake Engineering and Engineering Vibration* **5**, 119–131.

CEN (2004) *Eurocode 8: Design of structures for earthquake resistance – Part 1.5: Specific rules for concrete buildings*, EN 1998-1-5. European Committee for Standardization Brussels.

CEN (2006) *Eurocode 1: Actions on structures – Part 1.7: General actions – Accidental actions*, EN 1991-1-7. European Committee for Standardization, Brussels.

Corley, W.G., P.F. Mlakar, M.A. Sozen and C.H. Thornton (1998) The Oklahoma city bombing: summary and recommendations for multihazard mitigation. *Journal of Performance of Constructed Facilities* **12**, 100–112.

Dat, P.X. and T.K. Hai (2013) Membrane actions of RC slabs in mitigating progressive collapse of building structures. *Engineering Structures* **55**,107–115.

Department of Defense (2005) *Unified Facilities Criteria (UFC): Design of structures to resist progressive collapse*. DoD, Washington, DC.

Dusenberry, D.O. and G. Juneja (2003) *Review of existing guidelines and provisions related to progressive collapse*. Multihazard Mitigation Council of the National Institute of Building Standards, Washington, DC, pp. 1–31.

Ellingwood, B.R. (2006) Mitigating risk from abnormal loads and progressive collapse. *Journal of Performance of Constructed Facilities* **20**, 315–323.

Ellingwood, B.R. and E.V. Leyendecker (1978) Approaches for design against progressive collapse. *Journal of the Structural Division* **104**, 413–423.

Gross, J.L. and W. McGuire (1983) Progressive collapse resistant design. *Journal of Structural Engineering* **109**, 1–15.

GSA (2003) *Progressive collapse analysis and design guidelines for new federal office buildings and major modernization projects*. General Services Administration, Washington, DC.

Hartmann, D., M. Breidt, V. Nguyen, F. Stangenberg, S. Höhler, K. Schweizerhof, S. Mattern, G. Blankenhorn, B. Möller and M. Liebscher (2008) Structural collapse simulation under consideration of uncertainty – fundamental concept and results. *Computers and Structures* **86**, 2064–2078.

Hellesland, J. and A.C. Scordelis (1981) Analysis of RC bridge columns under imposed deformations. *IABSE Colloquium on Advanced Mechanics of Reinforced Concrete*, Delft, Holland, pp. 545–559.

Hughes, T.J.R. and W.K. Liu (1981) Nonlinear finite element analysis of shells: part II. Three-dimensional shells. *Computer Methods in Applied Mechanics and Engineering* **26**, 331–362.

Isobe, D. and M. Tsuda (2003) Seismic collapse analysis of reinforced concrete framed structures using the finite element method. *Earthquake Engineering and Structural Dynamics* **32**, 2027–2046.

Kaewkulchai, G. and E.B. Williamson (2004) Beam element formulation and solution procedure for dynamic progressive collapse analysis. *Computers and Structures* **82**, 639–651.

Khandelwal, K., S. El-Tawil and F. Sadek (2009) Progressive collapse analysis of seismically designed steel braced frames. *Journal of Constructional Steel Research* **65**, 699–708.

Kim, H.S., J. Kim and D.W. An (2009) Development of integrated system for progressive collapse analysis of building structures considering dynamic effects. *Advances in Engineering Software* **40**, 1–8.

Krauthammer, T. (1999) Blast-resistant structural concrete and steel connections. *International Journal of Impact Engineering* **22**, 887–910.

Kripakov, N.P., M.C. Sun and D.A. Donato (1995) ADINA applied toward simulation of progressive failure in underground mine structures. *Computers and Structures* **56**, 329–344.

Lee, C.H., S. Kim, K.H. Han and K. Lee (2009) Simplified nonlinear progressive collapse analysis of welded steel moment frames. *Journal of Constructional Steel Research* **65**, 1130–1137.

Li, Y., X.Z. Lu, H. Guan and L.P. Ye (2011) An improved tie force method for progressive collapse resistance design of reinforced concrete frame structures. *Engineering Structures* **33**, 2931–2942.

Ls-DYNA (2010) *User Manual*. Livermore Software Technology Corporation, Livermore, CA.

Lu, X., Y. Li, L. Ye and Y. Liang (2008) Application of fiber model for progressive collapse analysis of reinforced concrete frames. *Proceedings of 12th International Conference on Computing in Civil and Building Engineering*, Beijing, China.

Luccioni, B.M., R.D. Ambrosini and R.F. Danesi (2004) Analysis of building collapse under blast loads. *Engineering Structures* **26**, 63–71.

Mander, J.B., M.J.N. Priestley and R. Park (1988) Theoretical stress–strain model for confined concrete. *Journal of Structural Engineering* **114**, 1804–1826.

Marjanishvili, S.M. (2004) Progressive analysis procedure for progressive collapse. *Journal of Performance of Constructed Facilities* **18**, 79–85.

Menchel, K. (2009) *Progressive collapse: comparison of main standards, formulation and validation of new computational procedures*. PhD thesis, ULB Universite Libre de Bruxelles, Brussels, Belgium.

Möller, B., M. Liebscher, K. Schweizerhof, S. Mattern and G. Blankenhorn (2008) Structural collapse simulation under consideration of uncertainty – improvement of numerical efficiency. *Computers and Structures* **86**, 1875–1884.

Mpampatsikos, V., R. Nascimbene and L. Petrini (2008) A critical review of the R.C. frame existing building assessment procedure according to EUROCODE 8 and Italian Seismic Code. *Journal of Earthquake Engineering* **12**, 52–82.

Nair, R.S. (2006) Preventing disproportionate collapse. *Journal of Performance of Constructed Facilities* **20**, 309–314.

Nascimbene, R., G.A. Rassati and K.K. Wijesundara (2012) Numerical simulation of gusset-plate connections with rectangular hollow section shape brace under quasi-static cyclic loading. *Journal of Constructional Steel Research* **70**, 177–189.

National Research Council of Canada (1996) *National Building Code of Canada*. NRC, Ottawa.

NIST (2007) *Best practices for reducing the potential for progressive collapse in buildings*. United States National Institute of Standards and Technology, Technology Administration, US Department of Commerce, Washington, DC.

OpenSees – *Open System for Earthquake Engineering Simulation*, Pacific Earthquake Engineering Research Center, University of California, Berkeley, CA. Available from http://opensees.berkeley.edu (last accessed March 2013).

Park, J. and J. Kim (2010) Fragility analysis of steel moment frames with various seismic connections subjected to sudden loss of a column. *Engineering Structures* **32**, 1547–1555.

Paulay, T. and M.J.N. Priestley (1992) *Seismic design of reinforced concrete and masonry buildings*. John Wiley & Sons, New York.

Pretlove, A.J., M. Ramsden and A.G. Atkins (1991) Dynamic effects in progressive failure of structures. *International Journal of Impact Engineering* **11**, 539–546.

Priestley, M.J.N. and D.N. Grant (2005) Viscous damping in seismic design and analysis. *Journal of Earthquake Engineering* **9**, 229–255.

Saad, A., A. Said and Y. Tian (2008) Overview of progressive collapse analysis and retrofit techniques. *Proceedings of 5th International Engineering and Construction Conference*, Irvine, CA, pp. 765–772.

Santafé Iribarren, B., P. Berke, Ph. Bouillard, J. Vantomme and T.J. Massart (2011) Investigation of the influence of design and material parameters in the progressive collapse analysis of RC structures. *Engineering Structures* **33**, 2805–2820.

Sasani, M. (2008) Response of a reinforced concrete infilled-frame structure to removal of two adjacent columns. *Engineering Structures* **30**, 2478–2491.

Scott, M.H. and G.L. Fenves (2010) Krylov subspace accelerated Newton algorithm: application to dynamic progressive collapse simulation of frames. *Journal of Structural Engineering* **136**, 473–480.

Seismosoft. *SeismoStruct – A computer program for static and dynamic nonlinear analysis of framed structures*. Available from www.seismosoft.com (last accessed 2013).

Shi, Y., Z.-X. Li and H. Hao (2010) A new method for progressive collapse analysis of RC frames under blast loading. *Engineering Structures* **32**, 1691–1703.

Smyrou, E., C. Blandon, S. Antoniou, R. Pinho and F. Crisafulli (2011) Implementation and verification of a masonry panel model for nonlinear dynamic analysis of infilled RC frames. *Bulletin of Earthquake Engineering* **9**, 1519–1534.

Spacone, E., F.C. Filippou and F.F. Taucer (1996) Fibre beam-column model for non-linear analysis of RC frames: Part I. Formulation. *Earthquake Engineering and Structural Dynamics* **25**, 711–725.

Starossek, U. (2007) Typology of progressive collapse. *Engineering Structures* **29**, 2302–2307.

Starossek, U. (2009) *Progressive collapse of structures*. Thomas Telford, London.

Starossek, U. and M. Haberland (2010) Disproportionate collapse: terminology and procedures. *Journal of Performance of Constructed Facilities* **24**, 519–528.

Talaat, M.M. and K.M. Mosalam (2008) *Computational modeling of progressive collapse in reinforced concrete frame structures*, PEER Report 2007/10. University of California, Berkeley, CA.

Tsai, M.-H. and B.-H. Lin (2008) Investigation of progressive collapse resistance and inelastic response for an earthquake-resistant RC building subjected to column failure. *Engineering Structures* **30**, 3619–3628.

Usmani, A. S., Y.C. Chung and J.L. Torero (2003) How did the WTC towers collapse: a new theory. *Fire Safety Journal* **38**, 501–533.

Valipour, H.R. and S.J. Foster (2010) Finite element modelling of reinforced concrete framed structures including catenary action. *Computers and Structures* **88**, 529–538.

Vlassis, A.G., B.A. Izzuddin, A.Y. Elghazouli and D.A. Nethercot (2008) Progressive collapse of multi-storey buildings due to sudden column loss – Part II: Application. *Engineering Structures* **30**, 1424–1438.

Wei, J., R. Quintero, N. Galati and A. Nanni (2007) Failure modeling of bridge components subjected to blast loading – Part I: Strain rate-dependent damage model for concrete. *International Journal of Concrete Structures and Materials* **1**, 19–28.

Wibowo, H. and D.T. Lau (2009) Seismic progressive collapse: qualitative point of view. *Civil Engineering Dimension* **11**, 8–14.

Wijesundara, K.K., D. Bolognini, R. Nascimbene and G.M. Calvi (2009) Review of design parameters of concentrically braced frames with RHS shape braces. *Journal of Earthquake Engineering* **13**, 109–131.

Wijesundara, K.K., R. Nascimbene and T.J. Sullivan (2011) Equivalent viscous damping for steel concentrically braced frame structures. *Bulletin of Earthquake Engineering* **9**, 1535–1558.

Chapter 5
Use of calcium aluminate cements in sewer networks submitted to H₂S biogenic corrosion

François Saucier, Jean Herisson and Dominique Guinot

5.1 Introduction

Modern sewer networks are buried underground, and most people do not realise the size and complexity of this 'invisible grid' spreading under every single house in a city and down to sewage treatment plants. These infrastructures are subjected to a range of conditions, including soil movement, tree root infiltration and corrosion, and regular maintenance is required to ensure continuous service. One condition very specific to municipal sewers is 'H₂S biogenic corrosion' due to bacteria oxidising H₂S gas released by the effluent into sulfuric acid.

There are several options to address H₂S biogenic corrosion problems, either upstream, by impairing H₂S formation, or downstream, by venting out the H₂S or using building materials resistant to H₂S biogenic corrosion. However, even within a well-designed sewer network, a rule of thumb in the industry suggests that 5% of the total length may/will suffer from H₂S biogenic corrosion. In areas where favourable conditions exist, H₂S-related biogenic corrosion can deteriorate metal or several millimetres per year of concrete.

Because H₂S biogenic corrosion involves an acidic attack, one option for repairing corroded concrete infrastructures is to apply a lining of acid-proof material, such as an epoxy resin or other polymer-based liner. Although appealing at first sight, application of these hydrophobic and volatile-organic-compound-containing materials on confined, underground, moist concrete is a challenge. In the specific case of sewer infrastructures, too often the practical service life of such organic liners is below expectations. For this reason, greater attention has been paid recently to the biogenic corrosion resistance of calcium aluminate cement-based concretes and mortars to rehabilitate and/or protect sewer infrastructure.

This chapter explains why calcium aluminate concrete resists biogenic corrosion while classical concrete deteriorates. After a short summary of historical evidence of field performance, the key academic studies are presented, suggesting the benefit of using 100% calcium-aluminate-based mortar to maximise biogenic corrosion resistance. Finally, typical application methods, practical considerations and some references are presented.

5.2 Stakes

In most European countries, sewer networks are about 50 years old (on average) and need huge rehabilitation works; indeed, 10% of networks are more than 60 years old,

100

having been built for 60–80 years (Satin and Selmi, 2010). Concrete remains the most utilised building material. The size of these assets is quite astonishing. In the UK alone, the length of sewers servicing domestic, municipal and industrial uses totals 624,200 km, with about 96% of the population connected to a network. According to a 2009 study from the US Environmental Protection Agency (EPA), the expenses for 2009 in the USA for sewer rehabilitation were around US$3.3 billion (about 50% of the worldwide market).

With greater attention given today to water saving and quality, several highly visible mega-projects are being carried out (e.g. the Thames Tideway Scheme) to capture and direct all waste water to new sewerage treatment plants to reduce or eliminate the release of untreated water into rivers. However, the existence of such new projects should not hide the fact that the vast majority of the network is old and aging. Moreover, with water-saving campaigns, more efficient household appliances and the gradual separation of rain water and septic water, conditions have been becoming increasingly aggressive in some networks, and biogenic corrosion can be found today in locations where it has been absent for decades. The existing asset is just too big to be replaced, so owners of sewers have no option other than to rehabilitate corroding sewers, and choosing both a durable and affordable solution is a true challenge.

5.3 H₂S biogenic corrosion principles

The biogenic corrosion process in a septic sewer comprises a two-step ecosystem, as summarised in Figure 5.1. Municipal septic effluent by itself is not corrosive in the absence of industrial waste (its typical pH is around 7). However, it contains plenty of sulfated organic matter that can feed bacteria involved in the complex ecosystem leading to H₂S corrosion deterioration.

The first step takes place in locations where the effluent is deprived of oxygen, allowing some strains of anaerobic bacteria to use the sulfur as a source of energy by reducing sulfated organic matter. One result of this metabolic action is the production of hydrogen sulfide gas (H₂S). When transition times are long (gentle slope, slow flow of effluent), some anaerobic bacteria will produce enough H₂S to saturate the water and some H₂S gas will be released into the air above the effluent, especially in turbulent areas (e.g. at changes in flow direction in manholes, wet wells or pumping stations). Because the H₂S gas is heavier than air, it tends to stay within the sewer system.

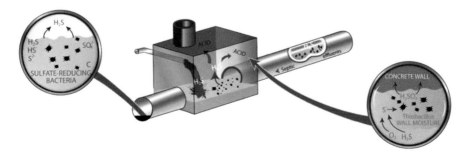

Figure 5.1 Principle of the biogenic corrosion ecosystem in a septic sewer.

The second step takes place in the aerial part of the sewer, where strains of aerobic bacteria develop on moist surfaces and use sulfur as a source of energy by oxidising it. The result of this oxidation is sulfuric acid (H_2SO_4). Different bacteria strains are able to oxidise sulfur, but the most deleterious for sewer infrastructures are the 'acidophilic' strains, which need acidic conditions to grow and develop.

New concrete surfaces have a natural pH of around 12–13. At this alkaline level, microorganisms cannot survive. However, through the action of atmospheric CO_2 over time, the concrete surface is gradually carbonated and the surface pH progressively decreases to 9. At this pH, some fungi and bacteria are able to develop. Each strain of bacteria has its own ideal conditions for growth, and they are able to modify their environment to get closer to these ideal conditions. When colonising concrete, the bacteria are trying to reduce the surface pH to below 9, as this level is quite aggressive for them.

Indeed, according to the ecosystem evolution model commonly proposed in the literature (Lamberet *et al.*, 2008; Figure 5.2), each strain of bacteria has its own preferred pH range for growth. *Starkeya novella* (formerly *Thiobacillus novellus*) is the first strain of bacteria to colonise the surface. It produces some acidic metabolites that begin to decrease the surface pH. This permits a second strain, *Halothiobacillus neapolitanus* (formerly *Thiobacillus neapolitanus*), to begin developing and lower the surface pH further. The third strain involved is *Thiomonas intermedia* (formerly *Thiobacillus intermedius*). When the surface pH reaches around 4–5, *Acidithiobacillus thiooxidans* (formerly *Thiobacillus thiooxidans*) can start to thrive. This bacterial strain, which was once named *Thiobacillus concretivorus* (Parker, 1947) because of its capacity to 'eat away' concrete, will produce sulfuric acid until the surface reaches a pH of 1–2, its ideal growth conditions.

This low pH is clearly well below the values that concrete and metal can withstand, and severe corrosion will occur. As long as acidophilic bacteria are supplied with nutrients, H_2S and moisture, they will produce acid. The reaction of sulfuric acid with ordinary Portland cement (OPC) concrete is dissolution of the cement paste and its reprecipitation as a soft and loose gypsum layer. Figure 5.3 illustrates how severe H_2S biogenic corrosion can be. The image shows an OPC concrete manhole in Egypt only 10 years after being put in service. If the corrosion rate is slower in less favourable conditions, the process remains the same.

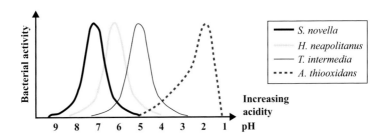

Figure 5.2 pH ranges for activity of bacteria of the former genus Thiobacilli (Lamberet *et al.*, 2008).

Figure 5.3 H$_2$S biogenic corrosion damage to an OPC concrete manhole in Egypt after 10 years in service.

It is interesting to note that H$_2$S biogenic corrosion is only a surface attack. The sulfuric acid from the biofilm is immediately neutralised on contact with the alkaline cement paste, so it cannot penetrate deeply into the cement paste. When the aggregates are of a siliceous nature (i.e. not sensitive to acidic conditions), only the cement paste is attacked by acid, leading to very uneven surfaces with protruding aggregates that will pull-out at some point. This acid attack mechanism is quite different from the well-studied concrete chemical attack by chloride ions or sulfates ions, which are able to penetrate within the porosity of the concrete to create later pathologies inside the concrete matrix. In fact, some authors suggest that sulfates from gypsum formed during the H$_2$S biogenic corrosion can get into solution at a later stage and then penetrate the concrete porosity to trigger further sulfate attack.

The aggressiveness of the H$_2$S biogenic corrosion ecosystem depends on several parameters, the most important being the following:

- *Presence of anaerobic areas*, allowing H$_2$S to form in the effluent. This condition happens easily in sediments accumulating at the bottom of the sewer. It also happens anywhere the sewer effluent remains stagnant long enough to become deprived of oxygen, for instance in a wet well, or even in long-pressure pumping pipes where the available oxygen is consumed by aerobic bacteria to the point that the effluent becomes anaerobic beyond a certain distance.
- *Concentration of the effluent.* As the waste water needs to be processed before ejection into the environment, the trend is to separate septic water from rain water to minimise the volume to be processed. This leads to more concentrated effluent with more organic 'nutrients' and less dissolved oxygen, allowing anaerobic conditions to develop more quickly. Reduction of infiltration waters also involves a reduction of the flow rate. Another impact of reduced flow is

that self-cleaning is less efficient, allowing more sedimentation and thus more H_2S-forming zones.

- *Long retention time.* As H_2S is formed only under anaerobic conditions, slow flow and long retention times give more time to aerobic bacteria to consume all available dissolved oxygen in water, creating anaerobic conditions. The flatter the land, the less slope in the sewer network, favouring slower flow, more pumping and longer retention time.
- *Favourable temperature for bacteria growth.* The temperature inside a sewer depends not only on the ambient climate but also on the temperature of the waste water, which is 'heated' by users (shower, washing, cooking, etc.). Even if a warm climate is a favourable parameter, the H_2S biogenic corrosion ecosystem also develops in countries with a cold climate.
- *Moisture.* Bacteria need water to develop and cannot thrive on a dry concrete surface. Inside a sewer, because the waste water temperature is generally warmer than the soil around the sewer infrastructure, condensation is quite likely to develop in most aerial parts, allowing biofilm to develop.

5.3.1 H_2S biogenic corrosion: a growing problem for owners

When designing a sewer, the engineer works the slopes, the diameters and the flow rates to minimise the transit time and to ensure good self-cleaning. Such a design aims to minimise, as much as possible, H_2S formation, and there are several reasons for this. For one, H_2S is a deadly gas and its presence complicates sewer maintenance by workers. H_2S is also characterised by a strong 'rotten eggs' smell, triggering complaints by citizens when it escapes the network through manholes and drains. Finally, H_2S 'feeds' the biogenic corrosion ecosystem, so the less H_2S that is available, the less corrosion will develop. For all these reasons, one might believe that H_2S biogenic corrosion of sewers is a well understood and mastered topic.

However, according to the authors' experience, H_2S biogenic corrosion, today, is a growing problem for numerous sewer operators in urban areas around the world. The same contributing factors seem to apply in many instances:

- First, the volume of water flowing in sewers is tending to reduce steadily over the years in sewers that were built to accommodate an increasing flow over time. One reason for this is the growth in water-saving campaigns. Citizens are reducing their water consumption either because of environmental considerations and/or in reaction to water-meter taxes. Recent household appliances have also become much more water efficient: showers are equipped with flow-reducing devices, washing machines and dishwashers use half the water they did 10 years ago. Also, because communities have to pay to clean waste water and work to avoid overflow due to rain, infiltrations are being fixed and new housing developments often separate septic and rain water into distinct networks. All these efforts are resulting in a decrease in the flow rate in sewers, which reduces self-curing; that is, it facilitates the sedimentation of sludge (where anaerobic bacteria form H_2S). All these factors are contributing to a greater production of H_2S. Lower flow rates also imply lower water levels, leaving a larger aerial part where acidophilic bacteria can thrive.

- A second reason arises from the 'urban sprawl'. In several cities, the H$_2$S issue was addressed for decades by natural ventilation of the sewer system, allowing H$_2$S to escape into the atmosphere rather than 'feeding' the bacteria. However, as cities grow, new houses are being built close to natural vents, and citizens complain of foul 'sewer odour'. The most immediate solution is to seal the vents. So, to solve odour complaints, H$_2$S is kept inside the sewer, increasing the H$_2$S biogenic corrosion potential.
- Another consequence of urban sprawl is the choice to close small local sewage treatment plants (STPs) and to collect and transport all the sewage water towards mega-STPs far from the city centre. This longer displacement of the sewage water often involves pumping in long pressurised mains where the water is rapidly deprived of oxygen, permitting the formation of H$_2$S during transport. When pumped water arrives at the release point, H$_2$S is released in quantity.

There are no two sewer systems that are identical, and these contributing factors combine in different ways in each location, but the same global trend seems to emerge in many countries: more severe H$_2$S biogenic corrosion is observed today than in previous decades.

One option to fix biogenic corrosion problems is to repair the corroding concrete with an acid-proof lining. Different polymers, like epoxy and polyurethane, can be sprayed or applied inside a corroding sewer to stop damage from the bacteria-borne sulfuric acid. Although these materials can resist pure sulfuric acid, they are hydrophobic materials and it is at a great challenge to obtain a flawless lining with durable bonding onto moist concrete. Applying the appropriate concrete-drying protocol in the harsh environment of a live sewer is often not possible, and ensuring an acid-proof membrane with not a single flaw requires a great deal of care. As a result, the rate of failure observed is too high to be acceptable.

As shown in Figure 5.4, typical polymer lining failure comes either from debonding or from the presence of flaws in the membrane (pin-holes, bubbles, cracks, breaks, leaking joints), permitting the ingress of bacteria-borne sulfuric acid. As the bacteria are able to lower the pH as low as 1, it is literally 'pure acid' coming in contact with the metal and concrete, creating rapid disruptive corrosion. The reaction of sulfuric acid with Portland cement paste leads to the formation of wet gypsum, which swells with enough pressure to burst the membrane.

There are other strategies to address H$_2$S biogenic corrosion beyond epoxy and inert linings:

- Chemical reactant like calcium nitrate can be continuously added to sewage water to impair the formation of H$_2$S.
- Active ventilation can be ensured by odour treatment units.
- Injection of compressed air into pressurised mains avoids the development of anaerobic conditions.

These three methods are efficient but they involve the continuous operation of mechanical equipment, which requires energy and consumable supplies.

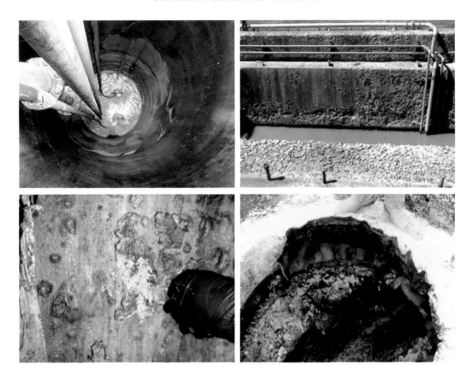

Figure 5.4 Example of polymer lining failures under H$_2$S biogenic corrosion conditions. Top left: debonding lining inside a pumping station (USA). Top right: damage to an epoxy-coated inlet channel (Australia). Bottom left: polymer lining perforated in many locations with swelling concrete underneath (Australia). Bottom right: polymer lining having failed to protect a manhole from H$_2$S corrosion (USA).

It is in that context that new attention is being paid today to the material being used. For example cementitious matrices made of calcium aluminate cement, are able to provide exceptional resistance to H$_2$S biogenic corrosion.

5.4 Portland cement provides limited response to H$_2$S biogenic deterioration

Ordinary Portland cements (OPCs) are mainly composed of tri-calcium silicate (C$_3$S: 50–70%), bi-calcium silicate (C$_2$S: 10–30%), tricalcium aluminate (C$_3$A: few %), calcium alumina-ferrite (C$_4$AF: 1–10%) and free lime (CH: 0.5–2%).

Once hydrated, the cement paste of OPC-based concrete or mortar is generally made of 50–60% calcium silica hydrate (C-S-H), 20–25% calcium hydroxide (CH), about 10% calcium sulfoaluminates, and various amounts of unhydrated cement grains (Mehta and Monteira, 1993). Acid decalcifies the hydrated products, starting with CH, which leads to weakening of the microstructure. Porosity increases and mechanical properties are reduced. After decalcification, calcium ions either leach out of the microstructure or combine with the acid salt to form insoluble calcium salts of little structural value (Pavlík, 1994; Duchesne and Bertron, 2013). This material is not durable in sewer environments prone to the biodeterioration phenomenon.

However, the cement industry now provides a large variety of materials by mixing OPC with various supplementary cementing materials (SCMs) that modify the properties of hardened concrete through hydraulic or pozzolanic activity. Typical examples of SCMs are natural pozzolans, calcined clays, ground granulated blast furnace slag, fly ash and silica fume (Peyre Lavigne *et al.*, 2016a). By co-grinding or mixing a single or different SCM with pure OPC clinker or cement, cement manufacturers produce a wide range of cement compositions (CEM I to CEM V in European standard EN 197-1).

The addition of an SCM to a pure OPC material leads to modification of the nature of the hydration products. Part of the lime produced during hydration of calcium silicates is consumed, and C-(A)-S-H and C-S-H with lower C/S ratios are formed (Juenger and Siddique, 2015). This leads to concrete or mortars with a lower level of porosity on a long-term basis and higher stability against aggressive environments. If a real gain in durability is noted with these materials, it is difficult to project this into gain in service life. Most studies investigating the effect of SCMs on resistance to sulfuric acid attack have been performed over short periods of time (from 28 days to 24 months; Atahan and Dikme, 2011; Ogawa *et al.*, 2012; Skaropoulou *et al.*, 2013; Nicolas *et al.*, 2014). Additional studies are thus needed to document the long-term behaviour of C-S-H and C-A-S-H bearing matrices facing H$_2$S biogenic acid attack.

5.5 Calcium aluminate resistance to H$_2$S biogenic corrosion: historical reminder

Calcium aluminate cements (CAC) alternatively named 'aluminous cements' or 'high-alumina cements' are cements in which calcium aluminate phases (CA, CA$_2$, C$_2$AS, C$_{12}$A$_7$ …) are the main constituents. Production involves two different manufacturing processes (sintering and fusion) by the reaction at high temperature of a source of lime (as limestone) and a source of alumina (as manufactured alumina or natural bauxite). Calcium aluminate cements are specialty materials used largely for making 'technical mortars' for the construction industry, in civil engineering applications, or in the refractory industry thanks to their ability to gain strength quickly or to withstand aggressive environments or high temperatures. However their H$_2$S biogenic corrosion resistance properties remain largely unknown. CAC was first patented in 1908 by Lafarge and sold under the brand 'Ciment Fondu®'. Today, a range of CACs are available, covering a broad spectrum of chemical compositions. Their chemistry and mineralogy differ from OPC, and this difference in composition gives quite different properties, which are often little known in the building industry (Hewlett, 2004; Newman and Choo, 2003). One of these properties is resistance to H$_2$S biogenic corrosion. The first mention of the use of CAC in sewers was in 1933 in Malaysia and Singapore.

Figure 5.5, left, shows ongoing repairs in a sewer in the Perth area of Australia, where Ciment Fondu® mortar is being applied as dry shotcrete inside a sewer to protect the underlying concrete. The estimated date is sometime in the 1950s. Although not much detailed information can be found for such old applications, Robson (1962) mentioned that CAC was regularly used to coat concrete pipes or to repair biogenic corrosion-damaged sewers.

Figure 5.5 Left: Sewer relined with gunited Ciment Fondu® in Perth (Australia) in 1950. Right: Precast concrete sewer pipe made in Durban (South Africa) in the 1960s.

Figure 5.6 Saint-Gobain Pont-à-Mousson (France) pipes (1977) for sewage transport, with a calcium aluminate mortar lining inside.

Figure 5.5, right, shows a precast concrete sewer pipe made in South Africa in the 1960s. The internal diameter is 1,350 mm and it is protected from H_2S biogenic corrosion by a 25 mm internal coating of Ciment Fondu® mortar. Around 30 km of such pipes were installed between 1955 and 1964 in the Durban area. An inspection conducted 31 years after the commissioning has shown no evidence of corrosion.

In 1977 the French ductile iron pipe maker Saint-Gobain Pont-à-Mousson decided to introduce pipes dedicated for sewage transport with a calcium aluminate mortar lining inside (Figure 5.6). This led to a new standard, and the current European Standard EN-598 only permits the use of calcium aluminate mortar or epoxy for the internal lining of sewer ductile iron pipes.

In 1991 a new step was made with the commercialisation of a 100% calcium aluminate mortar. It was introduced in the USA under the brand SewperCoat® as a repair material to rehabilitate sewer infrastructures severely damaged by H_2S biogenic

Figure 5.7 Raw materials used to create a 100% calcium aluminate mortar.

corrosion. Unlike previous sewer applications, where CAC was combined with natural aggregates, this new class of material was composed only of calcium aluminate (i.e. CAC combined with calcium aluminate aggregates to achieve a 100% calcium aluminate material). Figure 5.7 shows a typical calcium aluminate clinker block that is either milled to make cement or crushed to make aggregates. By combining these two fraction sizes, a unique type of repair material is obtained. The interest in this composition for higher biogenic corrosion resistance will be explained in the next section.

5.6 The science behind CAC resistance to H$_2$S biogenic corrosion

The academic study of H$_2$S biogenic corrosion is quite challenging because it is about observing and understanding a complex microscopic living ecosystem. Moreover, the interaction between this ecosystem and the building materials requires a multidisciplinary approach.

5.6.1 Hamburg simulation chamber

An important contribution was made by the team of Pr. Bock (Hamburg University) in the 1990s. This team developed a biogenic simulation chamber where the sewer ecosystem was reproduced for the testing of building material specimens in comparable conditions.

Figure 5.8 shows the principle of the Hamburg simulation chamber. It is a sealed volume of about 1 m^3, with the air saturated with moisture and the temperature maintained at 30°C. Concrete specimens are introduced into the chamber, and a variety of bacteria found in sewers are sprayed onto the specimens. H$_2$S is then fed in, along with nutrients for the bacteria. Even under ideal conditions it takes a few months to develop severely corroding conditions because the bacteria first need to colonise the surface to start decreasing the pH.

The Hamburg simulation chamber allows building materials to be exposed to a very severe but realistic cycle of H$_2$S biogenic corrosion. A typical test cycle lasts one

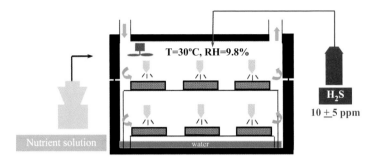

Figure 5.8 Schematic representation of the Hamburg chamber.

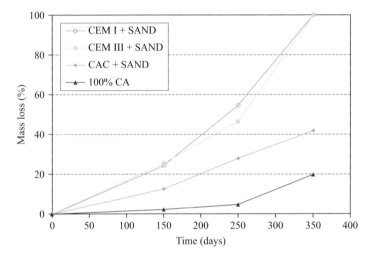

Figure 5.9 Weight loss over time of four cementitious materials exposed in the Hamburg simulation chamber for one year.

year. A comparison with companion specimens exposed in a live corroding sewer in the Hamburg area permitted correlation of that one year in the Hamburg simulation chamber with 24 years of exposure in the reference live sewer.

Figure 5.9 compares the weight loss over time for four series of mortar specimens exposed in the Hamburg chamber (Ehrich et al., 1999). The first two series, made of CEM I and CEM III, were completely corroded within one year in the simulation chamber. The third series, made of CAC with natural sand, showed a weight loss of around 45% after the same year of exposure. The fourth series, made of 100% calcium aluminate mortar (i.e. CAC combined with calcium aluminate aggregates) and exposed to the same conditions, showed only 20% weight loss after one year. This set of result shows that there is a drastic difference in behaviour between OPC and CAC concrete when exposed to the same H_2S deleterious environment.

In parallel with the monitoring of weight loss, specimen pH was also measured. Figure 5.10 shows the surface pH evolution over time for the same specimens as in

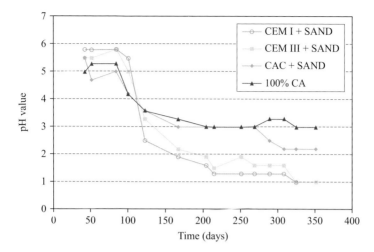

Figure 5.10 Surface pH evolution over time of four cementitious materials exposed in the Hamburg simulation chamber for one year.

Figure 5.9. It can be seen that for OPC specimens, bacterial activity lowers the pH to near 2 within six months, with a final pH of 1 after one year.

The behaviour on the CAC-based specimens is quite different. For those made with silica sand, a weight loss of 45% in one year was observed and the pH remained at 3 for a few months before decreasing to 2. For specimens made of 100% calcium aluminates (i.e. CAC combined with calcium aluminate aggregates), a weight loss of only 20% in one year was observed in the Hamburg simulation chamber, a pH of 3 was reached after six months, but it then remained stable over time. In fact, this stabilisation of the pH around a value of 3 is very characteristic of the capacity of calcium aluminate to interfere with the bacterial metabolism to stop the production of sulfuric acid. This mechanism is detailed hereafter.

The reason for the drastic difference between OPC and CAC in the face of H₂S biogenic corrosion lies in their different chemical compositions. With CAC, at least three different mechanisms contribute to the much better resistance to biogenic corrosion:

- The first barrier is the *larger neutralisation capacity* of CAC vs. OPC; 1 g of CAC can neutralise around 40% more acid than 1 g of OPC. This means that for a given production of acid by the biofilm, a CAC concrete will last longer. Moreover, when the concrete is 100% calcium aluminate, the whole matrix neutralises acid, not only the cement part (siliceous aggregates are inert to sulfuric acid). The higher neutralisation capacity is a positive factor for durability, but it does not explain fully the differences shown in Figures 5.9 and 5.10.
- The second barrier is due to the *precipitation of a layer of alumina gel* (AH_3 in cement chemistry notation) when the surface pH drops below 7–10 depending on the situation. At that pH, the calcium aluminate hydrates dissociates; on one side the calcium reacts with the sulfuric acid to form some gypsum, and on the other side aluminate precipitates as alumina gel. AH_3 is a stable compound

down to a pH of 4 and it will form an acid-resistant barrier as long as the surface pH is not lowered below 3–4 by bacterial activity.

- The third barrier is the *bacteriostatic effect* locally activated when the surface reaches a pH of less than 3–4. At this level, the AH_3 is no longer stable and will dissolve, liberating aluminium ions. These aluminium ions will accumulate in the thin biofilm. Once their concentration reaches 350–600 ppm, they will produce a bacteriostatic effect on the bacterial metabolism (i.e. bacteria will reach a dormancy state and will stop being active). In other word, bacteria will stop oxidising the sulfur from H_2S to produce more acid, and the pH will stop decreasing. This third barrier – the bacteriostatic effect – is the most important and it largely explains the calcium aluminate capacity to resist severe biogenic corrosion. Rather than trying to resist a continuous flow of acid, a calcium aluminate surface stops the production of acid, putting the corrosion process on hold.

This triple barrier mechanism does mean that some calcium aluminate hydrates are used up in the process, but the rate is slow enough to permit a long-lasting performance.

What Figures 5.9 and 5.10 are showing is that the mechanism is much more efficient when the whole surface presents a uniform chemistry (i.e. when both the cement paste and the aggregates are of the same chemistry). One possible explanation for this is that, on the scale of bacteria, sand grains (which do not liberate aluminium ions) are acting like 'virgin islands' where they can produce sulfuric acid without being slowed down by the bacteriostatic effect. The sulfuric acid produced on the sand grain surface can then run off on the surrounding cement paste and corrode it. As explained, calcium aluminate is not acid-proof, so if pure acid runs off, it will corrode the CAC, but because 100% CAC has the same chemistry throughout, there are no 'virgin islands' where the bacteria can produce sulfuric acid without being stopped when the pH reaches the 3–4 threshold value. It is this capacity to stop acid production that is key.

5.6.2 Strasbourg University study

A study run at the Strasbourg University in 2006–2007 (Geoffroy *et al.*, 2008) looked at the impact of aluminium ions on the metabolism of *Acidithiobacillus thiooxidans*. In fact, while aluminium ions are known for their metabolic effect for several bacteria strains, it was not established if *A. thiooxidans* was sensitive to this specific ion. Bacteria were grown in different culture media and then exposed to a large range of aluminium ion concentrations. With one culture medium, the threshold value to observe the bacteriostatic effect was around 350 ppm. With a second culture medium, the threshold was 600 ppm. This concentration can seem quite high but it must be kept in mind that the bacteria live at the surface of concrete in a very thin biofilm (not visible to the naked eye). Thus the quantity of aluminium ions required to reach the bacteriostatic threshold inside the biofilm remains very limited.

5.6.3 IFSTTAR study

From 2010 to 2013, a four-year research programme about H_2S biogenic corrosion in sewers was run at the French institute IFSTTAR (formerly the Laboratoire des

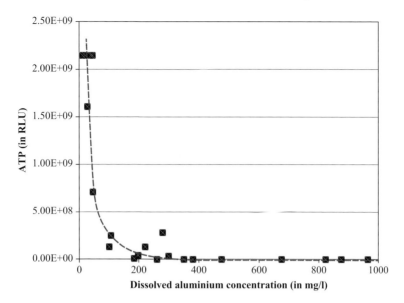

Figure 5.11 Bacterial activity in relation to total dissolved aluminium concentration.

Ponts et Chaussées (LCPC)). This provided new findings highlighting the difference between OPC and CAC exposed to biogenic corrosion (Herisson *et al.*, 2014a). For instance, abiotic tests (chemical tests without bacteria) were run to study H₂S affinity with cement. It has been shown (Herisson, 2012; Herisson *et al.*, 2013, 2014a; Grandclerc *et al.*, 2015) that the calcium aluminate surface has a lower chemical affinity with H₂S than the OPC surface and that H₂S is always absorbed faster on OPC mortars, whatever the initial H₂S concentration and the initial saturation rate of the specimens. A complementary study has been done about CAC surfaces by one of the authors to investigate the impact of the alumina content of the material on the reactivity to H₂S. This showed that the higher the alumina content, the less reactive to H₂S the material is. Thus, for a given time of exposure, less sulfur deposit forms on the CAC surface, and even less if the alumina content is high. Because it is the sulfur that bacteria oxidise, less sulfur results in less acid formation.

The IFSTTAR study also allowed additional measurements to be made on *in situ* specimens of the aluminium threshold that inhibits bacteria growth (Herisson, 2012; Herisson *et al.*, 2014a,b). The mortar specimens were hung in the sewer by hollow PVC tubes that become full of water from condensation, and this water was in contact with the mortar. Every four months specimens were inspected, and the condensed water inside the tubes was analysed for the activity of bacteria and aluminium ions content. Figure 5.11 shows this relationship between adenosine triphosphate (ATP) content – a tracker of bacteria activity – and aluminium content. With that data set, the threshold for bacteriostatic effect was around 350 ppm, confirming the results obtained by the study at Strasbourg University.

During the IFSTTAR study, specimens exposed to biogenic corrosion were observed with scanning electron microscopy (Herisson, 2012; Herisson *et al.*, 2014a)

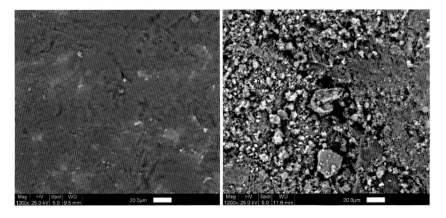

Figure 5.12 Scanning electron microscope images of specimens exposed to biogenic corrosion. Left: Surface of a 100% CA mortar smoothed by the AH[3] layer. Right: Surface of OPC mortar specimens showing a typical rough surface.

(Figure 5.12). This showed that calcium aluminate specimens were covered with a dense and smooth layer of a mineral compound that proved to be AH_3 once analysed. The presence of this AH_3 layer confirmed the mechanism described previously and allowed a better understanding of how the AH_3 can clog the porosity and reduce the penetration of aggressive compounds inside the cement matrix. Also, the smooth appearance of this layer led to a belief that it is not favourable for good bonding of the bacteria biofilm. Although the key role of the AH_3 layer is to cause the bacteriostatic effect, it probably participates in several ways to the overall excellent biogenic corrosion resistance.

5.7 The challenge of testing the biogenic corrosion resistance of building materials

Even if the H_2S biogenic corrosion process is well understood today, there is not yet a representative standard test method to give a realistic evaluation of the biogenic corrosion resistance of a given building material.

5.7.1 Pickle jar test method

In the literature and in industry, 'pure acid tests' predominate (Greenbook, 2009; ASTM, 2012), where specimens of materials are immersed in pure acid (e.g. sulfuric acid at pH 1) and weight loss is monitored over time. Such tests are often designed as 'pickle jar test'. These are simple to realise and are based on a simplistic assumption: 'because the pH inside a sewer can go as low as 1, let's test the material capacity to withstand a pH of one'. With a pure acid test, polymer materials show good resistance while cementitious materials deteriorate. These findings are thus extensively quoted by polymer material suppliers to prove the value of their products.

However, pure acid tests are misleading because they do not take into account the complexity of the biogenic corrosion ecosystem. As shown in Figures 5.9 and 5.10, calcium aluminate mortar has excellent resistance to biogenic corrosion. If the same specimens are submitted to a pure acid test at pH 1, they will gradually dissolve and will be rejected as non-resistant. To make sound choices, the design engineer needs standard test methods that are not only reproducible, but that provide a relative ranking representative of real-life behaviour.

To get closer to this double goal of 'reproducibility' and 'realistic ranking', several test methods have been proposed in the literature.

5.7.2 Heidelberg test method

The method proposed by Heidelberg University (Hormann *et al.*, 1997; Schmist *et al.*, 1997), despite the fact that it involved bacteria, was still a pure acid test. *A. thiooxidans* is cultivated in a fermenter at a pH of around 3.5. Specimens are then placed in a reactor and the bacteria solution is pumped over the specimens every hour, for five minutes. This procedure combined two aggressions. The first is a pure chemical attack, as specimens are immersed in a low-pH solution for five minutes per hour. As the volume of acidic solution is large, the alumina released under the acid attack cannot reach the threshold for the bacteriostatic effect. Between immersion cycles, bacteria are present on the specimen surfaces, but no equilibrium will be reached because the acidic nutrient solution is provided every 55 minutes. With this test procedure, it was shown that OPC mortars are ten times less resistant to biogenic deterioration than CAC mortars.

5.7.3 Gent University test method

Another attempt was made by Belgian researchers from Gent University (Vincke *et al.*, 1999; De Belie *et al.*, 2004). Their aim was to design a simple test combining the worst conditions found on site. One test cycle last 17 days and involves different steps: exposure to 250 ppm H₂S gas for three days; immersion in a solution enriched with *Thiobacillus* inside a rotary shaker for ten days; washing for two days; and finally drying for two days. Up to three cycles are completed before measuring mass loss and mechanical strength. Despite the fact that this test uses microorganisms, it was far from realistic conditions. Indeed, immersion for ten days in acidic culture medium is a pure acid attack. Moreover, the use of a drying episode in the cycle of deterioration limits the growth of microorganisms, so it impacts the acid concentration.

5.7.4 Virginia experimental sewer

In South Africa in the 1980s, the precast pipe manufacturing industry developed a very pragmatic approach to studying biogenic corrosion that remains unmatched today (Fourie, 2002; Alexander *et al.*, 2008). In the Virginia area, at a location known for its intense biogenic corrosion, a 65 m sewer was built with 900-mm-diameter industrial pipes made of nine different types of concrete. This section was provided with a bypass line allowing the water flow to be diverted on demand for regular inspection. The Virginia experimental sewer was commissioned in 1989. Inspections and measurements were done after 5, 12 and 14 years. At the 14th year, some OPC concrete sections were so badly corroded that they had to be removed before they collapsed. The removed sections were replaced by new set of specimens of shorter length, allowing new concrete compositions to be tested. What is very unique with this project

TABLE 15: MEASURED & ESTIMATED CORROSION RATES & MATERIAL FACTORS (11)

Cement/ Aggregate	5 year estimate		12 year estimate		14 year measured		Material factor***
	total	average	total	average	total	average	
PC/SIL	>30	>6,0	>64	>6,0	> 105	> 7.5	1.000
PC/DOL	10 – 15	2 – 3	20 – 30	1,7 – 2,5	43	3.1	0.410
CAC/SIL	5 – 10	1 – 2	10 – 15	0,8 – 1,2	26	1.9	0.250
FC	10 - 12	2 +	20 - 25	1,7 – 2,1			0.270
CAC/DOL *	3,0	0,6	7,2	0,6	8,4	0,6	0.085
CAC/ALM **							0.025

*Values estimated on the basis of other materials and performance of UCT samples in sewer
**Much less than CAC/DOL – no mass loss 17 months in sewer and pH on surface >6,4
***Average of maximum loss at side divided by corresponding value for PC/SIL.

Figure 5.13 View of Table 15 of the Sewer Design Manual (after Goyns, 2008).

is the exposure of full-sized industrially made precast concrete pipes in completely realistic conditions of a live sewer.

Using the data gathered with the Virginia experimental sewer, Goyns *et al.* (2008) proposed an improved version of the 'life factor method' (LFM) originally developed by Pomeroy and Parkhurst (1976). The original LFM only considered the alkalinity of concrete, but the Virginia experimental sewer results gave a sound base to take into consideration the whole influence of the material through a 'material factor' to be added into the LFM method. The South Africa pipe manufacturing industry published these results, the improved LFM and the material factor proposed value in a Sewer Design Manual published in 2008 (Goyns, 2008) (Figure 5.13).

In Sewer Design Manual Table 15, reproduced in Figure 5.13, the material factor proposed for conventional concrete based on OPC and siliceous aggregates (PC/SIL) is 1, meaning that the corrosion rate proposed by the Pomeroy equation is unchanged. However, if OPC is replaced with CAC (CAC/SIL), the material factor proposed is 0.25, meaning that the corrosion rate proposed by the Pomeroy equation needs to be divided by 4 to fit the real-life behaviour of CAC concrete pipes. In the same way, the material factor proposed for a concrete made with both CAC and calcium aluminate aggregates (CAC/ALM) is 0.025 (i.e. the expected corrosion rate will be 10 times lower than with only CAC concrete (CAC/SIL)). Although one can argue that a longer observation time would be required to define more accurately the material factor values, the fact is that calcium-aluminate-based concretes, exposed in a live sewer for up to 14 years, have shown a drastically different resistance to H_2S biogenic corrosion, in line with the Hamburg University findings.

There is also a difference between OPC and CAC exposed to biogenic deterioration in the deterioration products formed. The University of Cape Town (South Africa) worked on specimens from precast concrete pipes exposed for 10 years on site (Kiliswa, 2016). They initiated a microscopic study to describe the mechanism of deterioration for the various concrete mixtures exposed. According to their findings there are no cracks in CAC materials, but OPC specimens are deeply cracked due to the formation of disruptive ettringite. The same observation was made during other microscopic work done by one of the authors (Herisson *et al.*, 2016) on various

116

specimens exposed for four years in a French sewer network. Similarly, INSA Toulouse (Peyre Lavigne *et al.*, 2016b) has run some laboratory-scale tests with their own biodeterioration protocol to compare CEM III and CAC mortars made with siliceous sand. They also noticed the absence of ettringite in CAC material, with no cracks inside the matrix. These recent studies confirm that the presence of secondary ettringite is associated with the presence of cracks and high damage in OPC materials, while CAC materials are preserved because of its absence.

In 2016, 27 years after commissioning the Virginia experimental sewer, a new PhD work (Kiliswa, 2016) used a new set of data derived from segments of precast concrete pipes exposed for ten years on the site. One goal was to go a step further than the previously proposed material factor to improve predictions of the corrosion rate for various compositions of concrete, including CAC-based concrete. It was shown that introducing new parts in the equation brings enhanced results. $RC_{eff.g}$ is used instead of an alkalinity parameter. The first four elements of $RC_{eff.g}$ refer to the neutralisation capacity of the material and the last two to the bacteriostatic effect. It is expressed as follows:

$$RC_{eff.g} = \left(CaCO_{3(eq.,agg)}\right) + \left(CH_{(eq.,CAC)}\right) + \frac{\left(CH_{(eq.,PC)}^{-1}\right)}{100} + \left(AH_{3(eq.)}\right) + \left(Fe\left(OH\right)_{2(eq.)}\right) + \left(\frac{Al}{Ca}\right) + \left(\frac{Fe}{Ca}\right)$$

For CAC-based materials, input values for $CH_{(eq.,CAC)} > 0$ and those for $CH^{-1}_{(eq.,PC)} = 0$. For OPC-based materials, input values for $CH_{(eq.,CAC)} = 0$ and those for $CH^{-1}_{(eq.,PC)} > 0$.

Using this newly proposed equation, it was possible to match closely the measured corrosion rate of ten different concrete compositions exposed to H₂S corrosion conditions in the Virginia experimental sewer, including the concrete based on CAC. Further validation of this new equation may provide a new base for more accurate corrosion rate prediction for cementitious materials.

5.7.5 Fraunhofer UMSICHT simulation chamber

The Fraunhofer UMSICHT Institute came to the conclusion, more than 10 years ago, that the incidences of damage caused by biogenic sulfuric acid corrosion was so high that a testing method was necessary to select appropriate materials. In cooperation with and based on the work of Sand *et al.* (1994), they have developed a test rig consisting of different chambers allowing the testing of mortar, concrete, polymers and coated systems. This accelerated test is designed for industrial testing only.

The test principle is largely inspired by the Hamburg simulation chamber (Ehrich *et al.*, 1999) since Professor Sand was part of the Hamburg research team. The first four weeks of exposure is dedicated to the manual inoculation of specimens with a consortium of microorganisms coming from a biogas treatment plant and with specific microorganisms chosen for their ability to produce high amounts of sulfuric acid. The environment is kept warm (30°C), humid (100% relative humidity) and H₂S is introduced periodically in the form of pulses. A culture medium is regularly spread to feed microorganisms. Usually a test extends from nine months to a year during which visual evolution, weight measurement and surface pH are monitored.

A test conducted on a wide range of cementitious material led the authors to the conclusion that it is a representative test since the ranking of materials was similar to other on-site and laboratory studies.

5.7.6 Ongoing works to develop a biogenic corrosion standard test

Although providing highly credible real-life data, the Virginia experimental sewer programme is not a standard test method that can be reproduced elsewhere. Given the growing problem of H_2S biogenic corrosion in France, the French institute IFSTTAR started working in 2010 on a test method that could be candidate for a standard test method, based on the Hamburg simulation chamber. In the meantime, INSA Toulouse (Peyre Lavigne *et al.*, 2016b) has started to work on another test method with a slightly different approach. The results were interesting enough to trigger the launch in 2014 of a larger 4 million Euro four-year programme partly funded by the French government (called DURANET and involving main industry players) to improve the biogenic corrosion resistance of sewer materials. One expected result should be a test method that could form a reproducible, representative and accelerated 'standard test method'.

For the sewer owner who must choose the most appropriate materials to maintain/rehabilitate infrastructure, and considering the lack of a standard test method, field track records remain today the single most valuable source of information about the actual biogenic corrosion resistance of repair materials.

5.7.7 Available data concerning H_2S corrosion rate

Among the various research programmes on the H_2S biogenic corrosion behaviour of cementitious materials, only a few present data concerning the corrosion rate, either directly or by reporting mass loss. The corrosion rate is important for a designer to assess the potential service life of a given structure that will be exposed to H_2S biogenic corrosion. Table 5.1 summarises the reported corrosion rates – in mm per year – of three different studies: two made *in situ* in South Africa and in France, and one made in the laboratory in Germany. In the laboratory study, all parameters were optimised to accelerate the H_2S biogenic corrosion process inside the 'Hamburg simulation chamber', and the researchers evaluated an 'accelerating factor' of 24, meaning that specimens exposed inside the chamber corrode 24 times faster than companion specimens exposed in a real-life corroding sewer. The values in Table 5.1 are not directly comparable because, some were directly measured (Ehrich *et al.*, 1999; Kiliswa *et al.*, 2016) while others were calculated based on mass loss (ongoing French on-site study and Fraunhofer UMSICHT results, both unpublished). Likewise, the materials differ: some are mortars and others are concretes. Last, but not least, the environments vary in terms of temperature, relative humidity and H_2S concentration. Each column of Table 5.1 is a coherent set of data where comparisons are direct. More care should be taken when comparing the data of one column with the others.

Table 5.1 shows that, whatever the situation, the H_2S corrosion resistance of OPC-based materials is much less than the durability of CAC materials. As expected, the use of ground granulated blast furnace slag (GGBS) tends to enhance the durability of OPC materials but it reduces that of CAC (Herisson *et al.*, 2017). This is probably

Table 5.1 Corrosion rate reported in four studies.

	Accelerated test in laboratory		*In situ*	
	Hamburg biodeterioration chamber (Ehrich *et al.*, 1999)	Fraunhofer UMSICHT biodeterioration chamber (unpublished)	French onsite study (unpublished)	Virginia Experimental Sewer, South Africa (Kiliswa *et al.*, 2016)
	350 days	281 days	6.5 years	10 years
	Concrete	mortar	Mortar hung in aerial space	Concrete pipes in contact with sewage
Estimated accelerating factor	24	Not estimated	–	–
OPC + SIL	10.4 mm/y	16.5 mm/y	2.3 mm/y	–
OPC + DOL	–	–	–	1.5 mm/y
OPC + GGBS + SIL	–	–	1.8 mm/y	–
OPC + GGBS + DOL	–	–	–	1.3 mm/y
CAC + SIL	1.8 mm/y	4.6 mm/y	–	–
CAC + DOL	–	5.1 mm/y	–	0.5 mm/y
CAC + GGBS + SIL	–	–	0.88 mm/y	–
CAC + ALAG®	0.8 mm/y	3.0 mm/y	0.08 mm/y	0.2 mm/y

OPC: ordinary Portland cement; SIL: siliceous aggregates; DOL: dolomite aggregates; GGBS: ground granulated blast furnace slag; CAC: calcium aluminate cement; ALAG®: calcium aluminate aggregates.

due to the fact that GGBS contains more alumina than OPC but less than CAC. The use of acid-reactive aggregates (limestone aggregates, calcium aluminate aggregates) enhances the durability of the materials. Whatever the situation, the lowest corrosion rates are obtained with the 100% calcium aluminate materials (i.e. CAC combined with calcium aluminate aggregates).

The comparison between all these sets of data shows that it is possible to reproduce in the laboratory accelerated biogenic conditions able to differentiate cementitious materials with a ranking similar to what is observed in an actual sewer. Optimization of the living conditions for microorganisms (temperature, nutrient supply, humidity, etc.) is sufficient to produce significant acceleration of biogenic acid production onto materials surfaces.

These data confirm that the Pomeroy equation, which only takes into account the alkalinity of cement, underestimates the H₂S corrosion resistance of some cementitious materials, especially those containing CAC. For an engineer, always looking for a balance between cost and performance, knowing the potential corrosion rate of a given concrete permits the choice of materials and the design of a sewer infrastructure to be optimised.

5.7.8 Field evidence of 100% calcium aluminate mortar durability

Figures 5.9, 5.10 and 5.13 show published data underlining the large gap in durability between conventional OPC concrete and 100% calcium aluminate mortar/concrete.

119

Another valuable source of information is the monitoring over years of real-life reha-bilitation of 100% calcium aluminate mortars. Three different sets of data are pre-sented here.

5.7.8.1 Monterey Waste Water Treatment Plant

In California, USA, in the 1990s, the Monterey Waste Water Treatment Plant grid chamber was initially coated with an epoxy liner. Within 18 months, the epoxy liner showed severe distress and the concrete began to corrode badly. To choose a repair material able to withstand the severe corroding conditions, a benchmarking test was organised. Cylinder specimens (50 × 100 mm) were made with several candidate repair mortars, and were hung within the grit chamber in November 1995. The aver-age H_2S level was reported to be in the 30–50 ppm range for a period of six months. Visual observations and pH measurements were made after ten months and after two years. Although this study is empirical, the results are clear. Figure 5.14 shows the state, after two years of exposure, of a 100% calcium aluminate (CA) mortar specimen and of an OPC mortar specimen. The gap in performance is obvious, with the 100% CA mortar specimen showing almost no sign of deterioration after two years in an environment that was able to completely corrode a 50 mm OPC mortar cylinder.

5.7.8.2 Hampton Roads Sanitation District

The first rehabilitation with a 100% calcium aluminate mortar in USA took place in 1991 in Virginia, in a split chamber of the Hampton Roads Sanitation District. Very severe H_2S biogenic corrosion was ongoing, with 75 mm of concrete lost over a seven-year period. Surface pH measurements as low as 1.5 were reported. The chamber was rehabilitated by rebuilding the lost concrete with a dry shotcrete 100% calcium aluminate mortar. This repair was then monitored four times over a period of 11 years by visual inspection, hammer sounding and surface pH measurement.

The images in Figure 5.15 were taken during these inspections. On the left, the inspector is hammering the 100% CA mortar surface, which was found to be hard

Figure 5.14 In situ specimens exposed for two years in the grit chamber of Monterey Waste Water Treatment Plant. Left: 100% CAC mortar. Right: OPC mortar.

and sound. Note that the manhole cast-iron ring, which was protected with epoxy, is showing severe signs of corrosion. Figure 5.15, right, shows the pH measured after 11 years both on the cast-iron manhole hole ring and on the 100% CA mortar. On the epoxy surface, the pH is 1, showing that the biofilm has colonised the surface and is actively producing acid that has damaged the cast iron, despite the presence of an epoxy liner. On the 100% CA mortar, exposed to the same environment, the pH is 4, the same value that was recorded after three, six and nine years. This real-life observation illustrates well that the bacteriostatic effect of calcium aluminate has impaired the ability of the bacteria to lower the pH below 4 for over 11 years. At the same time, only a few centimetres away, the biofilm was allowed to thrive, without barrier, and lower the pH to 1. Note that the 100% CA mortar is not showing any sign of distress after 11 years is this harsh environment.

5.7.8.3 Ayer Rajah discharge chamber

In Singapore, the flat topography combined with the tropical climate is very favourable to H$_2$S biogenic corrosion in the sewer system. Figure 5.16, left, shows the state of decay of the Ayer Rajah discharge chamber in 2008, before its rehabilitation. The concrete is badly corroded, with rebars exposed and corroded. PUB, Singapore's national water agency, decided to rehabilitate this chamber using 100% calcium aluminate mortar. The other option would have been to bypass the flow for several weeks and build a new discharge chamber, but this would have been much more expensive and troublesome. Rehabilitation was carried out by rebuilding the missing concrete using a dry spray process in March 2008. A first inspection took place two years later, in August 2010, and the 100% CA mortar was found sound and solid, despite being stained by a yellowish compound, showing an active deposit of sulfur. On the base of this excellent performance, PUB authorised 100% CA mortar to be utilised for the regular maintenance of its sewer infrastructures. Since then, 100% CA technology has been utilised in Singapore.

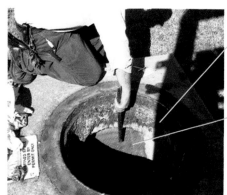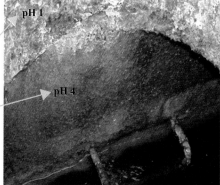

Figure 5.15 Manhole of Hampton Road Sanitation District rehabilitated with 100% CAC mortar. Left: Inspection of the rehabilitated surface. Right: pH measurement performed after 11 years in service.

Figure 5.16 Manhole of Ayer Rajah discharge chamber. Left: Corrosion damage before rehabilitation. Right: Application of 100% calcium aluminate mortar by a dry spray process.

5.8 Application methods and practical consideration

5.8.1 Main application of CAC in sewers

The H_2S biogenic corrosion resistance of CAC is utilised today in three main applications:

- *Ductile iron pipes* (DIPs) for sewage water, which have an internal lining made of CAC mortar
- *Concrete pipes* for sewers (seen mostly in Asia), which are made either with full mass CAC concrete or with an internal liner of CAC mortar
- *Rehabilitation* of man-accessible sewer infrastructures with 100% CA mortar

DIPs and concrete pipes are standardised products covered by national standards. Our focus here will thus be on practical consideration for rehabilitation work.

5.8.2 Application methods for 100% CA mortar

For rehabilitation works with 100% CA mortar there are three installation methods possible, depending on the job site constraints:

- *Low-pressure wet spray.* This method is the more common because it does not produce dust, and virtually no material is lost by rebound. It utilises a classical rotor pump, easily available in the market. Its main drawback is the limited pumping distance, which cannot exceed 75 m.
- *Spinning head wet spray.* This method is similar to the first, but manual spraying is replaced by a spinning head projecting the mortar onto the repaired surface. This method is fast and especially suited for cylindrical chambers like manholes. When a structure is so severely corroded that man entry is a risk, spinning head application permits an 'unmanned' consolidation of the manhole.

122

- *High-pressure dry spray.* This method, also called 'shotcrete' or 'gunite', allows a faster rate of rehabilitation and also a thicker application in a single pass. The main benefit of dry shotcrete is the capacity to pump the mortar over a long distance and this is needed when the access points are distant. The longest dry shotcrete distance the authors are aware of was at a job site in Australia where the 100% CA dry mortar was air transported over 800 m before being sprayed. The main drawback with dry shotcrete is the generation of dust and rebound; these could be limited with appropriate means such as a pre-moisture ring, adapted aggregate grading and an experienced nozzle man.

Whatever the application method chosen, the first step is a thorough cleaning of the corroded concrete to remove loose material and contaminants to expose a sound, rough and clean substrate. Depending of the concrete's condition and contamination, cleaning can range from simple high-pressure jet water cleaning (200 bar) up to hydrodemolition (2,000 bar). One method to ensure that sound concrete has been exposed is to verify that the pH is superior to 10. As for any concrete repair, state-of-the-art rules must be followed.

5.8.3 100% calcium aluminate mortar and conversion

When H₂S biogenic corrosion has removed several centimetres of concrete, the owner may want the rehabilitation not only to protect the structure from further deterioration, but also to restore structural capacity. As CAC is subject to the 'conversion phenomenon' (see below), this should be taken into account when structural capacity is required.

The 'conversion phenomenon' is specific to CAC: the metastable hydrates formed initially will transform over time into denser stable hydrates. Thus, the initial strength (one day to one month) may be higher than the long-term stable strength. This phenomenon is an inevitable thermodynamic process, so – for design purposes – only the long-term stable strength must be taken into account. As for other hydraulic cements, the strength of CAC is proportional to the water/cement ratio. The long-term stable strength can be evaluated by an accelerated test, as described in Annex 1 of EN 14647 Standard (2006).

In the specific case of 100% calcium aluminate mortar being utilised for water infrastructure rehabilitation, the water/cement ratio is typically moderate as a good stickiness on the walls or the ceiling of the repaired structures is required. With the positive effect of a moderate ratio and the good bonding between the calcium aluminate aggregate and the CAC paste, a long-term stable strength of 40 MPa is typically obtained.

5.8.3.1 Protection vs. consolidation

Depending on the progress of the H₂S biogenic corrosion damage, a given structure may only need to be protected from further damage, or it may need to be consolidated to ensure long-term stability. When only protection is needed, a 100% CA mortar can be applied with typical thickness ranging from 15 to 25 mm. When the H₂S corrosion has already removed several centimetres of concrete, the owner may want to rebuild the missing thickness to restore the structural capacity of the structure. Depending on the constraints specific to given job site, two choices are possible:

- Rebuilding the missing thickness with 100% CA mortar in a single pass (permitting a shorter intervention)
- Rebuilding the missing thickness with classical OPC concrete, and then protecting this conventional concrete with a protective layer of 100% CA mortar

The choice between these options really depends on the costs involved in a given location. Although 100% CA mortar is more expensive than plain OPC concrete, the capacity to make the rehabilitation in a single pass and to return to service within few hours may prove to be cheaper than having to intervene twice, once to install structural OPC concrete and then later to add the 100% CA mortar protection layer. The choice really depends on the constraints and cost structure of each repair.

5.8.3.2 Robustness to moisture

When a sewer needs to be rehabilitated, it is safe to assume that the underground concrete structure is either moist or water-saturated. Good bonding of the rehabilitation material must take this reality into account. With polymer liners like epoxy, because of the hydrophobic nature of these materials, the concrete surface needs to be dry enough to permit satisfactory bonding, and achieving surface dryness in a sewer is a challenge. In contrast, hydraulic cement-based mortars (like 100% CA mortars) need the concrete substrate to be either moist or even saturated surface dry (i.e. pores in aggregate are filled with water, but no surface film of water) to develop good bonding during hydration. Moreover, to ensure a good hydration of the mortar, moist curing is required and typically the ambient air in a sewer is quite humid. In other words, there is a 'good fit' with the moist environment of a sewer, and rehabilitation in a live sewer is possible with a 100% CA mortar. However, if the presence of H_2S requires forced ventilation while working, this may reduce the ambient air moisture, so active curing may be required to achieve good curing of the cementitious mortar.

5.8.3.3 Time for return to service

To proceed with sewer rehabilitation, it may be necessary to retain or divert flow during the repair. The cost of bypassing sewer flow is often the largest expense for a given operation. Thus the time to return to service after a repair is a non-negligible parameter. One property of CAC is a normal setting time followed by a fast strength development, within a few hours. Typical 100% CA mortar, which is formulated to ensure good pumpability and good bonding to the substrate, will start setting within 3–5 h, and can be put back into service after 8 h.

In some circumstances, the time available is shorter than that. For instance, for some large collectors, diverting the flow can be uneconomical, and rehabilitation instead takes place at night when the sewer flow is lowest. The working window can be as short as 6 h to clean the surface, proceed with the rehabilitation, let the material harden and return to service. When needed, it is possible to accelerate the 100% CA mortar hardening by applying a lithium-based accelerator. The hardening then starts within 30 minutes and the accelerated surface may support exposure to live water as soon as 1 h after the accelerator application. The application of a lithium-based accelerator is also a simple way to seal the 100% CA mortar surface to facilitate proper curing.

5.9 Conclusions

H$_2$S biogenic corrosion is a growing problem for sewer owners as the evolution in water usage tends to increase the H$_2$S production parameters. This chapter presents a broad review of the H$_2$S biogenic corrosion process and the rationale behind the exceptional durability of CAC in this deterioration process. The historical use of calcium aluminate in sewers is reviewed.

There are several possibilities for protecting or rehabilitating a sewer exposed to H$_2$S biogenic corrosion. The application of a polymer protection lining like epoxy is one option, but the failure rate observed over years by many operators is pushing them to search for an alternative solution with better reliability. Calcium aluminate mortar or concrete is such an alternative, its main weakness being that it is little known in the engineering community.

Conventional OPC concrete can be severely corroded by bacteria-borne sulfuric acid, but CAC resists much better because of its capacity to induce a bacteriostatic effect, stopping the production of acid by bacteria. It has been shown in academic studies and in field data that this resistance is maximised when the material is made entirely of calcium aluminates (i.e. both the cement and aggregates are made of calcium aluminate).

One missing element for the engineering community is a standard test method to permit the evaluation of concrete repairs materials resistance to H$_2$S biogenic corrosion. This chapter has shown why current methods like the 'pure acid test' do not provide realistic results. There is ongoing work towards achieving a standard test method proposal in the near future.

For the time being, actual field performance remains the most reliable source of data to evaluate the H$_2$S biogenic corrosion of repair materials. Today, there is a large body of evidence showing that calcium-aluminate-based materials present exceptional resistance to H$_2$S biogenic corrosion.

References

Alexander, M.G., A. Goyns and C. Fourie (2008) Experiences with a full-scale experimental sewer made with CAC and other cementitious binders in Virginia, South Africa. In *Calcium Aluminate Cement: Proceedings of the Centenary Conference*, Avignon, 30 June–2 July 2008, Fentiman C.H., R.J. Mangabhai and K.L. Scrivener, eds. IHS BRE Press, EP94, ISBN 978-1-84806-045-6, pp. 279–292.

ASTM C 267-01, *Standard test methods for chemical resistance of mortars, grouts, and monolithic surfacings and polymer concretes*. ASTM International, West Conshohocken, PA.

Atahan, H.N. and D. Dikme (2011) Use of mineral admixtures for enhanced resistance against sulfate attack. *Construction and Building Materials* **25**, 3450–3457.

De Belie, N., J. Monteny, A. Beeldens, E. Vincke, D. Van Gemert and W. Verstraete (2004) Experimental research and prediction of the effect of chemical and biogenic sulfuric acid on different types of commercially produced concrete sewer pipes. *Cement and Concrete Research* **34**, 2223–2236.

Duchesne, J. and A. Bertron (2013) Leaching of cementitious materials by pure water and strong acids (HCl and HNO_3). In *Performance of cement-based materials in aggressive aqueous environments*, Alexander, M. and N. De Belie, eds. Springer, Dordrecht, pp. 91–112.

Ehrich, S., L. Helard, R. Letourneaux, J. Willocq and E. Bock (1999) Biogenic and chemical sulfuric acid corrosion of mortars. *Journal of Materials in Civil Engineering* **11**, 340–344.

EN 197-1, *Cement – Part 1: Composition, specifications and conformity criteria for common cements*. BSI, London.

EN 14647, *Calcium aluminate cement – Composition, specifications and conformity criteria*, BS EN 14647:2005/AC, November 2006. BSI, London.

Fourie, C.W. (2002) *Biologically induced sulphuric acid attack on concrete samples in the experimental sewer section at Virginia*. Department of Civil Engineering, University of Cape Town.

Geoffroy, V.A., Bachelet, M., Crovisier, J.-L., Aouad, G. and Damidot, D. (2008) Evaluation of aluminium sensitivity on a biodegrading bacteria *Acidithiobacillus thiooxidans*: definition of a specific growth medium – calcium aluminate cements. *Proceedings of the Centenary Conference*, Avignon, 30 June–2 July 2008, Fentiman, C.H., R.J. Mangabhai and K.L. Scrivener, eds, EP94, pp. 309–319. IHS BRE Press, ISBN 978-1-84806-045-6.

Goyns, A.M. (2008) *Sewer design manual*. Pipe and Infrastructural Product Division, CMA, Midrand, South-Africa.

Goyns, A., M.G. Alexander and C. Fourie (2008) Applying experimental data to concrete sewer design and rehabilitation. In *Calcium Aluminate Cement: Proceedings of the Centenary Conference,* Avignon, 30 June–2 July 2008, Fentiman C.H., R.J. Mangabhai and K.L. Scrivener, eds. IHS BRE Press, EP94, ISBN 978-1-84806-045-6, pp. 293–308.

Grandclerc, A., M. Guéguen-Minerbe, T. Chaussadent and P. Dangla (2015) Etapes préalables à la mise en place d'un essai accéléré de biodétérioration des bétons dans les réseaux d'assainissement. *Proceedings of International Francophone Conference NoMaD 2015*, Mines Douai, France, p. 10.

Greenbook Standard Specifications for Public Works Construction (2009) Building News, Los Angeles, CA.

Herisson, J. (2012) *Biodétérioration des matériaux cimentaires dans les réseaux d'assainissement – Etude comparative du ciment d'aluminate de calcium et du ciment Portland (Biodeterioration of cementitious materials in sewers networks – comparative study of calcium aluminate cement and ordinary Portland cement)*, PhD thesis, Ifsttar, Université Paris-Est, Marne-la-Vallée, France.

Herisson, J., E.D. van Hullebusch, M. Moletta-Denat, P. Taquet and T. Chaussadent (2013) Toward an accelerated biodeterioration test to understand the behavior of Portland and calcium aluminate cementitious materials in sewer networks. *International Biodeterioration & Biodegradation* **84**, 236–243.

Herisson, J., E.D. van Hullebusch, M. Guéguen-Minerbe and T. Chaussadent, T. (2014a) Biogenic corrosion mechanism: study of parameters explaining calcium aluminate cement durability. In *Calcium aluminate: Proceedings of the International Conference*, Avignon, 18–21 May 2014, Fentiman C.H., R.J. Mangabhai and K.L. Scrivener, eds. IHS BRE Press, Bracknell, EP104, ISBN 978-1-84806-316-7, pp. 633–644.

Herisson, J., M. Guéguen-Minerbe, E.D. van Hullebusch and T. Chaussadent (2014b) Behaviour of different cementitious material formulations in sewer networks. *Water Science & Technology* **69**, 1502–1508.

Herisson, J., D. Tourlakis, S. Berger and H. Fryda (2016) *SEM observation of cementitious materials exposed to biodeterioration phenomenon*. Symposium of the RILEM TC253-MCI in Delft (The Netherland), June 2016.

Herisson, J., M. Guéguen-Minerbe, E.D. van Hullebusch and T. Chaussandent (2017) Influence of the binder on the behaviour of mortars exposed to H₂S in sewer networks: a long-term durability study. *Materials and Structures* **50**(8), doi:10.1617/s11527-016-0919-0.

Hewlett PC (2004) Calcium aluminate cement. In *Chemistry of Cement and Concrete*, 4th ed., Elsevier Science & Technology Books, Elsevier Butterworth-Heinmann, New York, Chapter 13, pp. 713–782.

Hormann, K., F. Hofmann and M. Schmidt, M. (1997) Stability of concrete against biogenic sulfuric acid corrosion, a new method for determination. In *Proceedings of the 10th International Congress on the Chemistry of Cement*, Gothenberg, 4vi038.

Juenger, M.C.G. and R. Siddique (2015) Recent advances in understanding the role of supplementary cementitious materials in concrete. *Cement and Concrete Research* **78**, 71–80.

Kiliswa, M.W. (2016) *Composition and microstructure of concrete mixtures subjected to biogenic acid corrosion and their role in corrosion prediction of concrete outfall sewers.* PhD thesis, University of Cape Town.

Lamberet, S., D. Guinot, E. Lempereur, J. Talley and C. Alt (2008) Field investigations of high performance calcium aluminate mortar for wastewater applications. In *Calcium aluminate cements. Proceedings of the Centenary Conference*, Avignon, 30 June–2 July 2008, Fentiman C.H., R.J. Mangabhai and K.L. Scrivener, eds. IHS BRE Press, Bracknell, EP94, ISBN 978-1-84806-045-6, pp. 269–277.

Mehta, P.K. and P.J. Monteiro (1993) *Concrete: Structure, properties and materials.* Prentice Hall, Englewood Cliffs, NJ.

Newman, B. and B.S. Choo (2003) Calcium aluminate cement. In *Advanced Concrete Technology: Constituent Materials Volume 1*, Elsevier Science & Technology Books, New York, Chapter 2, ISBN 9780750651035.

Nicolas, R.S., M. Cyr and G. Escadeillas (2014) Performance-based approach to durability of concrete containing flash-calcined metakaolin as cement replacement. *Construction and Building Materials* **55**, 313–322.

Ogawa, S., T. Nozaki, K. Yamada, H. Hirao and R.D. Hooton (2012) Improvement on sulfate resistance of blended cement with high alumina slag. *Cement and Concrete Research* **42**, 244–251.

Parker, C.D. (1947) Species of sulphur bacteria associated with the corrosion of concrete. *Nature* **159**, 439–440.

Pavlík, V. (1994) Corrosion of hardened cement paste by acetic and nitric acids part I: Calculation of corrosion depth. *Cement and Concrete Research* **24**, 551–562.

Peyre Lavigne, M., C. Lors, M. Valix, J. Herisson, E. Paul and A. Bertron (2016a) Microbial induced concrete deterioration in sewers environment: mechanisms and microbial populations. In *International RILEM Conference on Microorganisms–Cementitious Materials Interactions* eds. Wiktor, V., H. Jonkers and A. Bertron eds. RILEM Publications, Paris, pp. 20–36.

Peyre Lavigne, M., A. Bertron, C. Botanch, L. Auer, G. Hernandez-Raquet, A. Cockx, J.-N. Foussard, G. Escadeillas and E. Paul (2016b) Innovative approach to simulating the biodeterioration of industrial cementitious products in sewer environment. Part II: Validation on CAC and BFSC linings. *Cement and Concrete Research* **79**, 409–418.

Pomeroy, R.D. and J.D. Parkhurst (1976) *The forecasting of sulfide build-up rates in sewers.* Conference of International Association of Water Pollution Research, Sydney.

Robson, T.D. (1962) *High-Alumina Cements and Concretes.* Wiley, New York, pp. 145–159.

Sand, W., T. Dumas and S. Marcdargent (1994) Accelerated biogenic sulfuric-acid corrosion test for evaluating the performance of calcium-aluminate based concrete in sewage applications. In *Microbiologically influenced corrosion testing*, American Society for Testing and Materials, Philadelphia, Kearns J. R. Little, B. J. (Eds.), ASTM STP 1232, 234–249.

Satin, M. and B. Selmi (2010) *Guide technique de l'assainissement: Moniteur reference technique*. Edition du moniteur, Paris, ISBN 2-281-11239-X.

Schmidt, M., K. Hormann, F.J. Hofmann and E. Wagner (1997) Concrete with greater resistance to acid and to corrosion by biogenous sulfuric acid. *Concrete Precasting Plant and Technology* **4**, 64–70.

Skaropoulou, A., K. Sotiriadis, G. Kakali and S. Tsivilis (2013) Use of mineral admixtures to improve the resistance of limestone cement concrete against thaumasite form of sulfate attack. *Cement & Concrete Composites* **37**, 267–275.

Vincke, E., S. Verstichel, J. Monteny and W. Verstraete (1999) A new test procedure for biogenic sulfuric acid corrosion of concrete. *Biodegradation* **10**, 421–428.

128

Chapter 6
High cycle fatigue of concrete structures in harsh environments: design and monitoring

Mads K. Hovgaard

On a global scale, concrete is the preferred construction material by volume used. Many of the largest civil structures in the world rely on concrete for their main load-bearing systems, and it has also proved itself suitable for harsh environments, given its high durability and low maintenance costs. More than 40 concrete offshore structures have been commissioned in the North Sea, some of which are the largest structures ever moved across the surface of the Earth. More than half of all bridges in the USA are of a prestressed-concrete design.

This chapter focuses on the use of concrete for wind turbine towers, and on the topic of fatigue. Many wind turbine towers are in operation, most of which are in Germany, but they still constitute a small share of the total number of wind turbines. However, recent studies have found that the share of concrete wind turbine towers is increasing, and will reach 6% of offshore structures in 2016 (Gaspar, 2012). Sub-structures for offshore production facilities and wind turbines share some similarities, in that the loading is cyclic and that they are subject to strong performance require-ments. As concrete is a material of large inherent variability due to its method of construction, with multiple materials being mixed in a factory or on site, the impact of the environment at the time of production, the source and types of constituents, the type of cement, this variability also affects the physical parameters of the finished structure. The large scatter that is observed on small testing specimens in a laboratory can be expected to be larger on the actual structure, leading to difficulty in predict-ing properties. This in turn affects the level of safety of a structure, and this leads to higher partial safety factors (and a less economic design). The large variability in fatigue properties similarly causes fatigue safety factors to be very large, and this creates grounds for adapting a 'risk-based inspection' (RBI) framework, known from offshore engineering.

This chapter provides an in-depth introduction to concrete fatigue, with a focus on the empirical design models that have achieved inclusion in current standards. An overview of the physical damage phenomenon and state-of-the-art research into dam-age models is provided, as well as an introduction to monitoring technologies and to Bayesian decision analysis methods.

In the presented example, a wind turbine tower is considered and the expected costs with regard to fatigue failure are considered. Primarily, a probabilistic model for fatigue reliability is established by combining results from the literature. By incor-porating structural health monitoring technology, based on novelty detection of the

129

measured global dynamic response of the structure, and using a correlation model between Miner's fatigue damage index D and the measured secant modulus of the concrete within the affected region, Monte Carlo sampling is performed to estimate the expected costs, finally enabling calculation of the value of information (VoI). The VoI is the primary indicator of the utilitarian value of an inspection strategy or a monitoring system.

The result of the numerical investigation is that the monitoring system has a negative benefit. This is mainly because of the very high uncertainties related to the damage-sensitive feature model. On the other hand, sensitivity analysis of the fatigue reliability showed great potential for 'usage monitoring', as the probability of failure's largest sensitivity is to the model uncertainties on the loads.

6.1 Introduction and background

The interest in fatigue in reinforced concrete peaked in the 1970s with the introduction of concrete in offshore structures. The oil industry operating in the North Sea used post-tensioned concrete as a construction material, with the first gravity-base platform Ekofisk installed in 1973. The UK followed two years later with Beryl Alpha (Ocean Structures, 2009), and over the following three decades a further 46 concrete structures were installed in the North Sea. Offshore structures are subject to high-cycle fatigue loading from waves, and are designed to withstand approximately 10^8 cycles during their service life. The large application scope, as well as the high economic risks associated with these assets, has spurred a renewed interest in fatigue resistance in harsh environments (Shah, 1982).

When fatigue became important for concrete, the first approach was to fit the laws of steel fatigue to the outcome of experiments. Unlike steel, concrete has different strength properties in tension and in compression, and this introduced the need for a modification to the fatigue design laws for steel (known as the Wöhler curves after the German railway engineer who discovered them). The current design model is still based on a modification of the Wöhler curve. Concrete fatigue gained considerable interest with the design and construction of the first Norwegian gravity based Condeep offshore platforms in the 1970s (Figure 6.1) (Holmen, 1984) but, although many codes (CEB-FIB, 1993; fib, 2012; EC, 2004; DNV, 2012) incorporate rules for concrete fatigue design, few to no failures have been observed. In a report from 2009 (Ocean Structures, 2009), 27 Condeep structures were investigated, but no cases of concrete fatigue were diagnosed. Concrete fatigue has been under investigation for causing excessive creep of several long-span box-girder bridges, including the collapsed Palau bridge (Bazant and Hubler, 2014) and recent full-scale tests in northern Germany have attempted to provoke fatigue in a gravity base foundation (Urban et al., 2014). The importance of fatigue for future structures, as discussed by RILEM Committee 36 (1984), is becoming increasingly relevant.

Fatigue of concrete or steel is inherently very uncertain, and large scatter is observed in testing specimens (Cornelissen and Reinhardt, 1982; Lassen, 1997; Breitenbücher and Ibuk, 2006). This causes predictive models for fatigue to be associated with large uncertainties, and the safety factors to be correspondingly large. For critical components of jackets and topsides in offshore engineering, an expected 'time to failure' $E[t_f]$ of up to 10 times the service life is not uncommon, 10 being the 'fatigue

Figure 6.1 A Condeep substructure being rowed out (attributed to O. Furenes).

design factor'. Fatigue is also the primary design driver for many structural components of wind turbines, including the blades, tower, foundation and hub.

6.1.1 Background to inspection and monitoring of structures

Reliability with regard to fatigue failure can be updated by the use of information from the often expensive manual inspections (Madsen *et al.*, 1986), the advantage thereof being a more economic design. As a supplement or alternative to inspections, a monitoring system, consisting of various sensors positioned on the structure, can provide information. The RBI discipline uses the information from inspections, or similar activities such as monitoring, to update the predicted reliability of a structure, using a framework of Bayesian decision analysis (Raiffa and Schlaifer, 1961). The optimisation strategy is maximisation of the expected utility, by co-optimising inspection strategies and structural parameters. The premise of such analysis is the existence of probability of detection (PoD) models, which describe the performance of the inspection method. To enable such PoD models, it is the primary concern to define the type of damage manifestation that is targeted, and this is where things get complicated for concrete. Although steel fatigue is dominated by the growth of a single macroscale crack, concrete fatigue seems to cover different damage types, corresponding to either tensile or compressive loading. In pure tension, a crack growth similar to that of steel has been observed, but for compression, the deterioration seems to be the increase of a volumetric region of microcracks. Damage such as the latter cannot be observed visually, so other technologies must be used. Inspection of reinforced concrete can be visual or using non-destructive testing equipment, such as ultrasound or acoustic emission technologies. These technologies are all categorised

131

Table 6.1 Low-level comparison of inspection and monitoring technologies.

	Inspections	Structural health monitoring (SHM)	Usage monitoring (UM)
Examples of technology	• Visual • Eddy current • Ultrasound • Acoustic emission	• Vibration-based (global) • Guided waves (local)	• Strain gauges (local) • Vibration-based (global)
Pros	• Provides information of structural health • High detection rate and low rate of false positives	• Provides information of structural health • Information is obtained constantly • Sensors can also provide information for UM	• Low installation cost and low to no maintenance costs • Information is obtained constantly
Cons	• High costs for each instance	• Less precise than inspection • High rate of false positives • High installation cost, as system must be tailored to each application • Maintenance costs for entire life of structure	• Only provides information of load effects

as 'local' inspection technologies, in the sense that they can only be applied to a single point at a time, and the equipment must be relocated to cover all points of interest on the structure.

Another approach to obtaining information about the structure is to use monitoring technology. Monitoring technologies fall into two categories: usage monitoring (UM) and structural health monitoring (SHM) (Table 6.1). Rather than the sparse information of inspections, SHM provides vast amounts of information. However, as the sensors of an SHM system can only provide information that correlates to the state of the structure, rather than actual measurements of damage size, it can be expected to perform at a much lower precision.

There is no current RBI approach for concrete fatigue. The structures are designed according to the 'safe-life' concept, where the reliability against fatigue failure is sufficient for the whole service life, without inclusion of inspection data. In some applications, this leads to a very conservative design.

6.2 Stress–life theory of fatigue

Fatigue has been the known cause of many structural failures since it was first recognised and methodically tested for conveyor chains by Albert in 1838. Throughout the 19th century it was observed in components and structures for transportation, and it has gained considerable attention following several catastrophic failures attributed to fatigue, including the Versailles train crash (1842), the Liberty ships (1943), the Comet airplanes (1954) and the Alexander L. Kielland offshore platform (1980).

Fatigue is the gradual deterioration of a material due to repeated occurrences of stress. It was initially generally agreed that fatigue is driven by cyclic loading, but it was not until the middle of the 20th century that it was generally agreed that fatigue

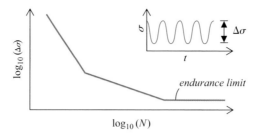

Figure 6.2 Wöhler curve S–N model with endurance limit for constant-amplitude loading. The slope of the curve is called the Basquin slope.

is a local damage mechanism, rather than a gradual deterioration of the material as a whole. Its manifestation in metals and other homogenous materials is a very local three-stage crack-growth process, consisting of 1) nucleation, 2) propagation and 3) fracture.

In 1860 the German railway engineer August Wöhler conducted fatigue tests. By counting the number of cycles to failure N_f for various stress ranges $\Delta\sigma$ of metallic specimens, he fitted curves in a double-logarithmic N_f and $\Delta\sigma$ coordinate system. He also observed that for stress amplitudes below a certain endurance limit, failure would not occur (Wöhler, 1860). Half a century later, in 1910, O.H. Basquin of Northwestern University formulated the empirical law of metal fatigue (Basquin, 1910), based on idealised Wöhler curves. The theory is known as the S–N, or 'stress–life' theory. An S–N model is sketched in Figure 6.2.

The scope of Basquin's equation, which is based on constant-amplitude tests, was later extended to include variable loading by Palmgren (1924) and Miner (1945) through the concept of 'linear damage accumulation'. The combined theory enables fatigue design, and, to this date still forms the basis of fatigue design for all materials. Although the theory can be used to predict failure, it is strictly empirical and provides no explanation for the physical mechanisms that cause failure.

The phenomenon in steel is the progression of a single macrosized crack, and in 1963 the 'linear elastic fracture mechanics' (LEFM) model of crack propagation was published by Paris and Erdogan (1963). The Paris law only models the stage 2 part of the growth, where the macroscale crack has formed and is propagating.

6.2.1 Stress–life theory for metals

Metal fatigue has been the focus of thorough research, and several empirical laws have achieved inclusion in international standards. Fatigue failure in structural steels is a brittle failure type that often occurs with little or no premonition for operators. According to Lassen and Recho (2006), fatigue initiates from microscopic defects in locations with large stress concentrations. Both these factors are present in welds, and fatigue often initiates there. The environment influences fatigue life, and the presence of corrosion reduces fatigue life. Lassen and Recho provide a thorough overview of the S–N approach, which is briefly introduced in the following.

Fatigue loading is categorised into three regimes: low-cycle (<10^4 cycles), high-cycle (>10^4 cycles) and ultra-high cycle (>10^8 cycles). The low-cycle regime, which is observed on bending a paper clip until it breaks, is mainly dominated by nonlinear behaviour and large plastic strains. The forces that cause such effects are inherent to catastrophic events (e.g. earthquakes) where very violent actions cause repeated yielding and plastic deformation. By definition, all loads that are not static are time-varying. However, normal structures subject to wind loading are not considered cyclic-loaded structures. It is more often the resonance (i.e. dynamic amplification), initiated by variations in the loading, that causes fatigue. High-cycle loading is the normal design regime of the structural codes and several empirical stress–life models have achieved inclusion in international standards. They are based on the Basquin (1910) equation:

$$\log_{10}(N) = \log_{10}(K) - m\log_{10}(\Delta\sigma) \quad \leftrightarrow \quad N = K\Delta\sigma^{-m} \tag{6.1}$$

where N is the number of cycles to failure, $\Delta\sigma$ is the stress range, and K and m are empirical constants. Most experiments have been performed with fewer than 10^7 cycles. Offshore structures undergo in the range of 10^8 load cycles from waves, and structural components of wind turbines endure in the region of 10^9 cycles and thus fall into an area where knowledge of fatigue behaviour is very limited. The design codes are extrapolated into the ultra-high cycle region, and some codes implement an 'endurance limit' rather than extrapolate into the regime. The model describes the fatigue damage as a function of stress range alone. Corrections have been suggested to account for the influence of a non-zero mean stress, but these are not discussed here. The values for constants m and K depend on the type of stress calculation concept used, on the type of alloy, on the surface treatment and on the environmental conditions of the detail, and may be given as 5% fractiles for deterministic analysis. For variable-amplitude stresses, the linear damage accumulation hypothesis by Palmgren–Miner (Miner, 1945) is used to calculate a damage sum D:

$$D = \sum_{i=1}^{k} \frac{n_i}{N_i} \tag{6.2}$$

where n_i is the number of cycles and N_i is the number of cycles to failure for stress range i. Palmgren–Miner's model assumes that the order of the load effects has no influence on the fatigue life. Some S–N models are bilinear. The limit state function is then

$$g = \Delta - \sum_{i=1}^{k} \frac{n_i}{N_i}, \; N_i = K_j(\Delta\sigma_i)^{-m_j}, \; j = \begin{cases} 1 \; if \; \Delta\sigma_i > \left(\frac{K_2}{K_1}\right)^{(m_2-m_1)^{-1}} \\ 2, \; \text{otherwise} \end{cases} \tag{6.3}$$

where $\Delta\sigma_i$ is the damage sum at failure. Some models also have an endurance limit (i.e. a stress level $\Delta\sigma_{cutoff}$), and stresses less than or equal to this do not contribute to fatigue life.

The uncertainties in the S–N model approach are both of the epistemic (model and measurement) type and of the aleatory (inherent) type. According to Straub (2004), they can be categorised into load, model and resistance uncertainties. The uncertainties may be represented by random variables, but, due to the nature of the underlying experiments, each variable cannot be isolated for statistical analysis and some engineering judgement must be applied.

6.2.2 Stress–life theory for concrete and other cementitious materials

Fatigue in reinforced concrete can mean fatigue of the concrete or fatigue of the reinforcement. In the following, only fatigue of the concrete is considered, as reinforcement fatigue would only be relevant if large tensile strains occur and the concrete fatigue capacity would then already be exhausted. Unlike steel, concrete is a quasi-brittle cementitious, heterogeneous material with largely different properties in tension and in compression.

Fatigue design of concrete structures is (although there are some differences from the design of steel structures) based on Wöhler-type S–N curves and Miner-sum linear damage accumulation. For metals, Basquin's equation describes cycles to failure for a given stress range, but for concrete, the fatigue life is based on two parameters: minimum and maximum relative stress S_{min} and S_{max}. The ratio of minimum to maximum stress is called the R ratio. Due to the different tensile and compressive properties of concrete, there are three regimes:

- Pure compression ($1 < R < \infty$)
- Pure tensile ($0 \leq R < 1$)
- Alternating cycles ($-\infty < R < 0$)

Many influencing factors necessitate modification of the laws known from metal fatigue. The more pronounced modifications are the mean stress level and the loading sequence in the application of linear damage accumulation for variable amplitude loading (Cornelissen, 1984).

The majority of tests are carried out under constant-amplitude loading, and only a few tests have used variable-amplitude loading. The lack of data makes the use of the Palmgren–Miner hypothesis to account for variable-amplitude and random loading the most obvious choice. However, as Holmen (1979) has shown, loading history has an effect on fatigue life. Holmen observed that reversing the order of his loading sequence would adversely affect the results of cycles to failure. The results showed that small stresses followed by larger stresses resulted in a shorter life of the specimen. In the case of random loading, which is characteristic for structures loaded by wind or waves over a longer span of time, it seems reasonable to assume linear damage accumulation, as the load cycles are randomly ordered.

6.2.2.1 Compressive regime, $1 < R < \infty$

The evolution of the S–N models started in 1970, when Aas-Jakobsen (1970) proposed the following adapted Wöhler curve for pure compression:

$$S_{max} = 1 - \beta_{aj} \cdot (1 - R) \cdot \log_{10}(N) \qquad (6.4)$$

where the slope of the Wöhler curve β is an experimentally decided constant and R is the ratio of minimum stress to maximum stress. He performed multiple tests, under

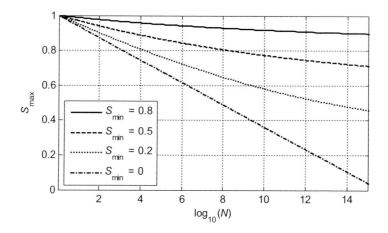

Figure 6.3 Aas-Jakobsen's Wöhler curves for pure compression.

constant-amplitude loading, the results of which provided the basis of 0, but also using alternating loading.

Aas-Jakobsen's expression is shown in Figure 6.3. Aas-Jakobsen suggested that $\beta = 0.064$. Although most expressions that have since been proposed are based on equation 6.1, the β factors differ slightly. Tepfers and Kutti (1979) proposed $\beta = 0.0685$ and RILEM (1984) suggest values between 0.064 and 0.080. In the recent fib model code 2010, the curves were modified to the bilinear model shown in Figure 6.8.

Stemland *et al.* (1990) observed that the number of cycles to failure increased at load cycles above 10^6 cycles and thus proposed a bilinear model with endurance limit, here rearranged in a similar form to 0 (Figure 6.4):

$$S_{max} = 1 - \frac{1}{8 \cdot S_{min}^2 + 16 \cdot S_{min+12}} \cdot \log_{10}(N), \ for \ N \le 10^6$$

$$S_{max} = 1 - \frac{1}{\left(\log_{10}(N) + 5\right)\left(8 \cdot S_{min}^2 + 16 \cdot S_{min+12}\right)} \cdot \log_{10}(N), \ for \ N > 10^6 \qquad (6.5)$$

For $N < 10^6$, the slope of 0 varies from 0.028 to 0.083.

6.2.2.2 Tensile and alternating regime, $-\infty < R < 1$

Tepfers (1979) stated that the modification by Tepfers and Kutti (1979) also describes tensile fatigue well, as long as the stresses are scaled to the static strengths, respectively, and that alternating stresses do not cause faster deterioration than pure compressive or tensile cycles with $R = 0$.

Cornelissen (1984) showed that the stress distribution in the specimen had a large impact on the cycles to failure. This implies that direct axial (concentric) tests will yield lower cycles to failure, as opposed to the flexural test, where stress redistribution may occur. The finding influences the choice of experimental data to include in the same empirical model, and in turn also makes the choice of model applications

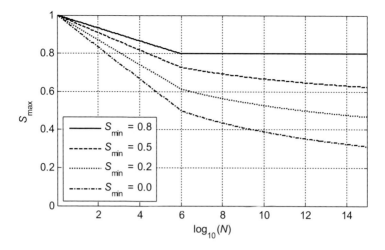

Figure 6.4 Wöhler curves after Stemland *et al.* (1990) for pure compression.

specific. Under the assumption that flexural alternating tests may lead to higher fatigue strengths due to redistribution of stress, Cornelissen performed concentric tests in tension and in alternating stress. The outcome led to the conclusion that alternating stress reduces the fatigue life more than tensile or compressive cycles alone. The following expression relates to alternating tension–compression:

$$S_{max} = 1.18 - 0.323 \cdot S_{min} - 0.126 \cdot \log_{10}(N) \qquad (6.6)$$

For tension with a lower limit of 0, the following expression was found:

$$S_{max} = 1.02 - 0.0689 \cdot \log_{10}(N) \qquad (6.7)$$

Both expressions are shown in Figure 6.5.

6.2.2.3 Design method in the current codes

Most current design codes are based on Aas-Jakobsen's relation for compressive cycles, with minor modifications. In the recent model code 2010 (fib, 2012), a bilinear model is used. This was developed from the model presented by Stemland *et al.* (1990). The latest modification was motivated by recent results in ultra-high-strength concrete experimental trials.

The model in CEB model code 90 (CEB, 1993) is based partly on the model by Stemland *et al.* (1990) and Petković *et al.* (1990), with the same initial slope (Figure 6.6).

The current Eurocode EN 1992-1-1 (European Committee for Standardization, 2004) bases its verification on $\beta = 0.072$, which becomes apparent after rewriting to the following form:

$$S_{max} = 1 - 0.072\sqrt{1 - R} \cdot \log_{10} N \qquad (6.8)$$

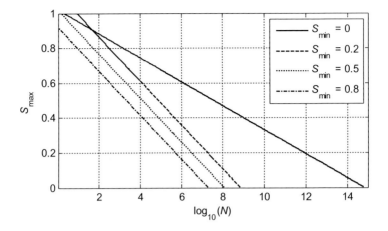

Figure 6.5 Cornelissen (1984) Wöhler curves for tension and in alternating stress for concentric specimens.

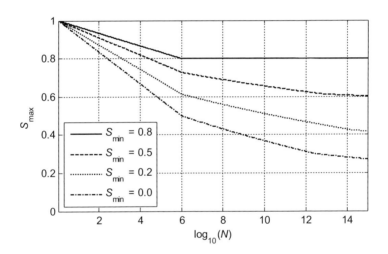

Figure 6.6 CEB (1993) model code 90 Wöhler curves for pure compression.

The EC2 curves are shown in Figure 6.7.

The deduction of model code 2010 (fib, 2012) expression for cycles to failure of concrete in compression is given in Lohaus *et al.* (2012) and in Wefer (2010). The modification was motivated by recent results from ultra-high-strength concrete experimental trials. The model code 2010 expression is shown in Figure 6.8.

6.2.2.4 Probabilistic stress–life analysis method

The need for a probabilistic model for the fatigue life of concrete was promoted by Oh (1986) and McCall (1958), for example. Among the motivating reasons are the following:

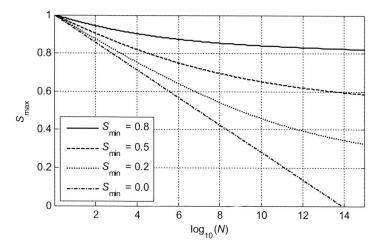

Figure 6.7 Eurocode 2 EN 1992-1-1 (European Committee for Standardization, 2004) Wöhler curves for pure compression.

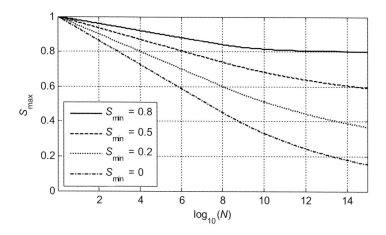

Figure 6.8 fib model code 2010 Wöhler curves for pure compression (fib, 2012).

- The large material uncertainties of concrete create a very large scatter on fatigue strength, and the Palmgren–Miner hypothesis is erroneous for variable-amplitude loading (Holmen, 1979).
- Tests that use different material properties are gathered under the same model, and the scale effects are very large (Bazant and Xu, 1991).
- Very few tests have been carried out in the alternating regime, which is relevant for partially post-tensioned concrete structures.

Oh (1986) reasoned that probabilistic methods can be applied to better account for the stochastic nature of concrete fatigue. Furthermore, SHM-based structural design is only possible through knowledge of the conditional probabilities of events, and probabilistic analysis enables these probabilities to be assessed. However, a probabilistic approach, making use of level 2 methods of structural reliability (Thoft-Christensen and Baker, 1982), requires knowledge of the joint distributions of all variables relating to the problem. These distributions are hardly obtainable and some simplifications, such as assumptions about distributions types and the correlation of variables, are required.

A probabilistic formulation of the model in Figure 6.8 from model code 2010 (fib, 2012) is chosen to approximate the limit state function for compressive fatigue. The model represents the state of the art and is based on regression on experimental data. This gives it a broad range of validity. For alternating stresses, model 0 from Cornelissen (1984) is used. The choice of this model is based on the fact that the underlying experiments are concentric with alternating stress. The elimination of the stress gradients gives similarity to the stress distribution in the wall of a wind turbine tower. As both models are deterministic models containing parameters fitted to experimental data, subjective model uncertainties were added. The high scatter of fatigue life was observed by Holmen (1979), Cornelissen and Reinhardt (1982) and Cornelissen (1984) to be linked to the uncertainty model for f_c. To account for conditions different from those in the performed tests, model uncertainties for the compressive and tensile regimes $X_{N,c}$ and $X_{N,t}$ respectively are added. The model uncertainties are chosen so that the model for $\log_{10}N$, in accordance with Oh (1986), remains Weibull distributed. The uncertainty model is shown in Table 6.2. See Appendix A for MATLAB code for probabilistic model of concrete fatigue.

The uncertainty models, especially concerning correlations of f_c, f_{ct} and X_{Nt} and X_{Nc}, remain a topic for future research. As the same concentric specimens have not, for obvious reasons, been tested both statically and dynamically, additional sources of uncertainty cannot yet be identified or quantified. In a probabilistic analysis, prior reasoning as well as some conservatism must be applied in setting the model

Table 6.2 Uncertainty model used for reliability analysis of the concrete wind turbine tower.

Variable	Type	Mean	Standard deviation	Comment
μ_σ, $\Delta\sigma$, n	Deterministic			Rainflow counted load cycle matrix
f_c	Log Normal	f_{cm}	$0.15f_{cm}$, $\rho = 0.9$	From JCSS (2006)
f_{ct}		$0.3f_{cm}^{(2/3)}$	$0.3f_{ctm}$	
F_p	Normal	F_{pm}	$0.05F_{pm}$	Post-tensioning force
X_{Nc}	Log Normal	1	0.016	Subjective model uncertainty: compressive
X_{Nt}	Log Normal	1	0.037	Subjective model uncertainty: tensile
Δ	Log Normal	1	0.3	Model uncertainty on Palmgren–Miner, from Holmen (1979)
X_s	Log Normal	1	0.132	Combined LN load uncertainty, from IEC61400-1 (IEC, 2005)
X_{aero}	Gumbel	1	0.1	Load uncertainty, from IEC61400-1 (IEC, 2005)

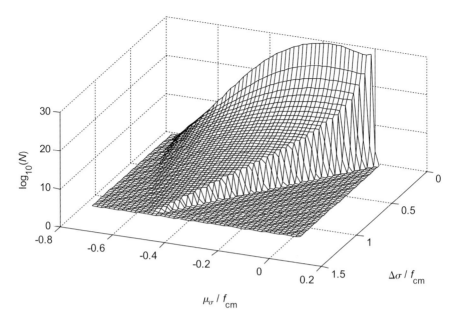

Figure 6.9 Fatigue model: expected cycles to failure as a function of mean-stress μ_σ/f_{cm} and stress range Δ_σ/f_{cm}.

uncertainties. The probability of failure at the tower foot is shown in Figure 6.13. The model uncertainties for the loads have been included in Table 6.2. The expected value of the cycles to failure $E[N_f]$ forms the surface plotted in Figure 6.9.

6.2.2.5 Example: concrete tower for the NREL 5 MW wind turbine

A post-tensioned, slip-formed concrete tower for the Jonkman *et al.* (2009) NREL 5 MW baseline onshore wind turbine is used as a design case. The turbine is modelled in aero-elastic simulation software LACflex 1.6.2. Wind conditions are modelled as wind class II-B with directional conditions corresponding to a Danish coastal area. Simulated load effects are extracted as cycle count matrices and extrapolated to a service life of 20 years. The extrapolated load matrix for the tower foot is shown in Figure 6.10.

The fatigue reliability of the concrete tower was first estimated using a 'first order reliability method' (FORM), and the estimates were verified by importance sampling with the code FERUM 4.1 (Der Kiureghian *et al.*, 2006; Bourinet *et al.*, 2009). Dual design points necessitate a system-level analysis, and a FORM series system approximation was used to satisfying accuracy, as verified with Monte Carlo sampling. The accumulated failure probability for the concrete tower is shown in Figure 6.11.

The annual reliability index β_{annual} is seen to be above $\beta_{min} = 3.0$, corresponding to a suitable component reliability for a wind turbine (Sørensen and Toft, 2010). The shape of the accumulated probability of failure corresponds to the results of Petryna and Krätzig (2005). It can be observed that a substantial contribution to the expected failure costs fall within the first 5% of the service life. From the distribution of cycles to failure for the concrete tower shown in Figure 6.11 it can be deduced that while the

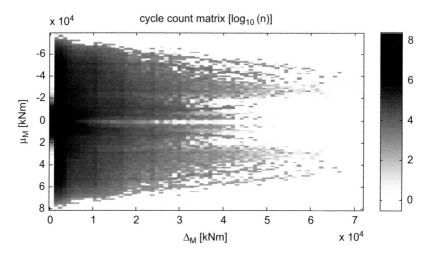

Figure 6.10 Graphical representation of the cycle count matrix of resulting bending moments at the tower foot used in the fatigue design.

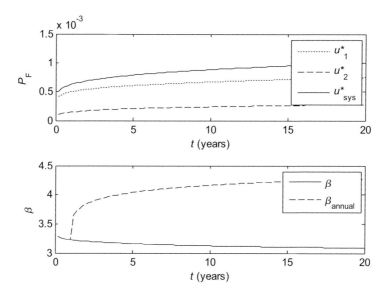

Figure 6.11 Fatigue reliability of the original (prior) structure without SHM. Top: Probability of failure with contribution from the two design points shown. Bottom: Reliability index.

first fraction of the service life is dominated by tensile fatigue, the main part is dominated by compressive fatigue. This is valuable knowledge, as it means that the damage that should be targeted by inspections and monitoring should be related to compressive fatigue, rather than tensile fatigue. Specifically for monitoring, this means that the damage-sensitive feature model must have a high sensitivity to the damage incurred by compressive fatigue. This is further elaborated in the following section.

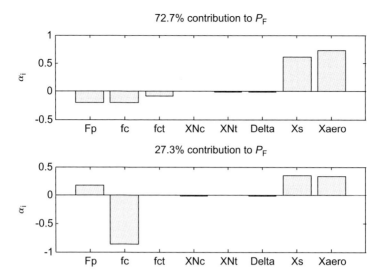

Figure 6.12 Sensitivity analysis: α-vectors from the two design points of FORM analysis at $t = 20$ years.

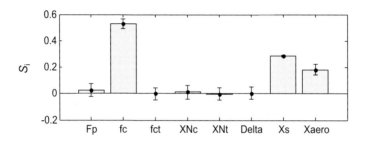

Figure 6.13 Sobol global sensitivity indices with 95% confidence bounds.

The alpha vectors from the FORM analysis provide a sensitivity analysis of the reliability index, as shown in Figure 6.12. It can be observed that the two load model uncertainties X_s and X_{aero}, representing uncertainty on exposure (terrain), climate statistics, shape factor, structural dynamics and stress evaluation (all taken from the uncertainty model proposed in IEC, 2005), as well as cylinder strength f_c, have a large impact on reliability.

A global sensitivity analysis is, unlike the above FORM sensitivity analysis, not based on the design point but on the full parameter space of basic variables. According to Sobol (2001), it is suitable to evaluate the sensitivity of the model, rather than the sensitivity at a specific solution. The result of a global sensitivity analysis is shown in Figure 6.13.

6.3 Physical damage mechanism

For a steel structure, fatigue manifestation is the well-defined crack growth, and the cracks are targeted by inspections or SHM. For concrete, knowledge of the fatigue mechanics is much more limited. Fatigue experiments with concrete specimens are usually concentric tests on cylinders or cubes, or flexural tests on small beams. The size effect, when compared to a large civil structure, is pronounced, as the microscale damage growth that causes failure of small unreinforced specimens may well not dominate the progression in full-scale structures. It is plausible that a macroscale damage progression constitutes the main part of the fatigue progression, and that microcracking has the same role in concrete fatigue as crack-nucleation in steel fatigue. However, very little research has gone beyond small-scale tests, and thus the basis for design is still based on small-scale tests, which we shall also assume in the following.

Fatigue growth in concrete is a progressive growth of cracking. Under reversed or tensile loading, cracks interact and join, causing accelerated deterioration (Cornelissen and Reinhardt, 1982). The damage causes an increase in load-produced strains that follows three stages: an initial period of decrease of the strain-increase rate, a longer period of constant rate, and a final period of increasing rate of strain increase. This three-stage progression is analogous to the phenomenon of crack growth in metals, for which the second stage can be described by the Paris law (Paris and Erdogan, 1963):

$$\frac{da}{dn} = C\Delta K^m \qquad (6.9)$$

The Paris law uses the LEFM similarity model, in which crack propagation is caused by stress intensity factor K at the crack tip. The value of K drives stable crack growth, and fracture occurs when K exceeds fracture strength Kc. As concrete has a heterogeneous microstructure, the LEFM model used to describe fatigue crack propagation in metals is inapplicable. In search of an FM model that could be used for concrete, the 'fictitious' crack model was developed by Hillerborg et al. (1976). The difference from the normal model is that the crack tip is replaced with a process zone called the 'fracture process zone'. The Paris law cannot be directly applied to model fatigue damage in concrete, as the mechanism is not the same. Where steel fatigue is characterised by growth of a single macroscale crack, the deterioration of concrete is driven by mechanisms at the microscale over a larger process zone (Bazant and Hubler, 2014). At all length scales, the structure of concrete is disordered and defects such as microcracks and macrocracks are both present. Under reversed or tensile loading, the cracks interact and join, causing accelerated deterioration (Cornelissen and Reinhardt, 1982).

In the early 1970s Holmen (1979) showed that the secant modulus E_s of the concrete decreased during compressive fatigue loading and that the size of the reduction increased with reduced minimum stress level S_{min}. Holmen used high stress levels, not likely to be found in civil structures ($S_{max} > 0.675$). The phenomenon that Holmen observed is known as 'cyclic creep', which was first observed by Feret (1906). A mathematical model was recently developed using fracture mechanics on the microscopic crack growth, both in tension and compression, by Bazant and Hubler (2014). This model is, however, valid only at the microlevel and is generally too involved to model a propagation of damage at the macrolevel, which is required for SHM purposes. The deterioration causes an increase in load-produced strains that follows three stages: 1) an initial period of decrease of the strain-increase rate, 2) a longer period of constant rate, and 3) a final period of increasing rate of strain increase (Figure 6.14).

144

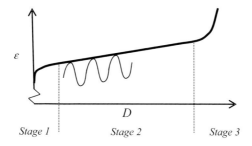

Figure 6.14 Principle of cyclic creep.

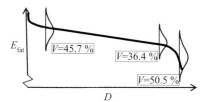

Figure 6.15 Sketch of the development of the secant modulus E_{fat} as a function of the theoretical Miner damage sum. Values of the coefficient of variation V are taken from Breitenbücher and Ibuk (2006).

The secondary stage constitutes the larger part of the fatigue life and shows a correlation, albeit weak, to the deterioration state. The currently used non-destructive evaluation (NDE) methods for concrete fatigue assessment are ultrasound and acoustic emission, as described in Urban *et al.* (2014). They target the level of deterioration based on empirical correlations to the damage sum D, through Holmen's cyclic creep relation.

The uncertainty model shown in Figure 6.15 is for compressive cycles ($S_{min} = 0.05$, $S_{max} = 0.675$) based on research from 2006 at the Ruhr University, and is presented by Breitenbücher and Ibuk (2006). Whereas Holmen in 1979 conducted his tests at an S_{max} stress level above 0.675, these newer tests use lower stress levels, and are more representative of the stress conditions in civil structures. More recent research by Thiele (2015) investigates several damage-sensitive features, including the secant modulus and ultrasound velocity. Thiele makes the important observation that the different values of the secant modulus, corresponding to different parts of the work curve, have very different damage sensitivities. The three values included were $E_{f,1/}$, $E_{f,m}$ and $E_{f,max}$, corresponding to the part from unloaded $S \sim 0$ to $S = 1/3$, the part from minimum cyclic stress S_{min} to mean level S_m, and the part from mean level S_m to maximum cyclic stress S_{max}. Of these, $E_{f,1/3}$ showed the highest damage correlation combined with best noise rejection (Figure 6.16).

The secant modulus relation, where the fatigue damage state is correlated to the secant modulus, is valuable in SHM applications, as it suggests that the secant modulus E_s is a damage-sensitive feature (Figure 6.17). Recent research in Germany by Grünberg and Göhlmann (2006) uses Holmen's model to calculate damage levels.

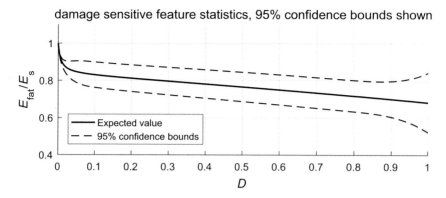

Figure 6.16 Evolution of the low-strain dynamic secant modulus $E_{f,1/3}$ as a function of theoretical damage sum D. Results from Thiele (2015).

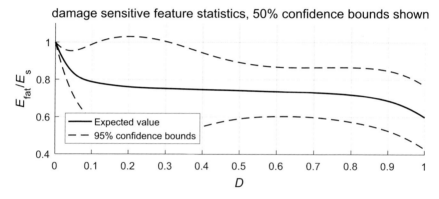

Figure 6.17 Evolution of the secant modulus E_{fat} as a function of theoretical damage sum D. Results from Holmen (1979) and Breitenbücher and Ibuk (2006).

Urban *et al.* (2014) use the model as predictive model for a full-scale fatigue test of a wind turbine gravity base foundation, using acoustic emission and ultrasonic SHM techniques to detect the deterioration state. However, to the authors' knowledge, no results have yet been published from these investigations.

6.4 Inclusion of information from inspections and monitoring

SHM is the process of making operations and maintenance (O&M) decisions based on real-time information about damage-sensitive features. To account for SHM in a design, Bayesian decision analysis (Raiffa and Schlaifer, 1961) is used. An optimal detector can be designed by minimising the expected loss with respect to system variables and then selecting an O&M decision strategy, given knowledge of the detector performance (Flynn, 2010; Hovgaard and Brincker, 2016).

Figure 6.18 Sensing and decision process flowchart.

If the stress–life analysis can be correlated to a damage-sensitive feature, such as Eigen frequencies or vibration level, through stiffness reduction, then Bayesian methods can be used to update the fatigue reliability, given information from inspections or SHM. The flowchart in Figure 6.18 shows the sensing and decision process.

The field of SHM has many branches, all of which can be categorised into global monitoring or localised monitoring. Global methods refer to global physical parameters, such as modal parameters. The measurable modal parameters (frequencies, mode shapes, damping) are functions of the physical properties of the structure (stiffness, mass, damping). The premise of vibration-based SHM is that damage from deterioration cause changes in physical properties (Farrar and Worden, 2012) and that these changes can be detected and identified though corresponding changes in the modal properties. However, local methods are the natural starting point for developing new SHM technologies, as these methods are not subject to the same uncertainties of location and damage type as the global methods. Local methods are directed at measurements of a single location (or smaller area), which can be selected as hot spots, that is, fatigue-critical details and locations. Examples of local methods applied to concrete include the ultrasonic pulse velocity technique and acoustic emission (RILEM, 1984). The advantage of global methods is that larger civil engineering structures, such as wind turbine towers, would require vast sensor arrays to monitor the whole structure using localised methods. For example, local methods would perhaps require a sensor for every square metre of the surface, resulting in several hundreds of sensors, whereas a global system would require perhaps only 30 sensors. This makes global methods the only feasible option. The downside of global methods is inferior performance; as mechanical excitation of large civil engineering structures for an experimental modal analysis is intractable, due to the magnitude of force required for excitation, operational modal analysis (OMA) is the only tractable method. The uncertainties in the OMA extracted modal parameters are large for many reasons (e.g. lack of stationarity in the environmental conditions, lack of broadband excitation and bias from the identification algorithm). These large uncertainties adversely affect the performance of the monitoring system, and risk-based decision-making is required to ensure valuable SHM performance. In the following, Bayesian decision analysis methods are introduced and the wind turbine tower example is then continued.

6.4.1 Introduction to Bayesian decision analysis

Bayesian decision analysis in an SHM context is introduced in the following.

Lindley (1971) uses the framework of Bayesian decision analysis to present 'Bayesian experimental design', as summarised in the following. For any design $e \in E$, data $x \in X$ are observed and decisions $d \in D$ are made. The unknown parameters are $\theta \in \Theta$. The sequence is visualised by the decision tree in Figure 6.19.

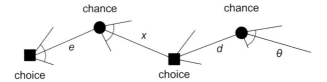

Figure 6.19 Decision tree representing the decision sequence forming the basis of Bayesian experimental design.

The expected posterior utility of the best decision is found by

$$\max_{d} \int_{\theta} U(e,x,d,\theta)\, p(\theta|x,e)\, p(x|e)\, d\theta \qquad (6.10)$$

This is an optimisation of posterior utility as $p(\theta|x)$ is conditioned on the data. It should be stressed here that e represent all design variables of the system (e.g. choice of design variables, choice of experiments, choice of inspections). By setting an appropriate utility function U, optimisation of decisions (e, d) yields the optimal design, found by averaging over data x and optimising:

$$\max_{e} \int_{x} \max_{d} \int_{\theta} U(e,x,d,\theta)\, p(\theta|x,e)\, p(x|e)\, d\theta\, dx \qquad (6.11)$$

This expression is a pre-posterior decision analysis and an unconstrained dual-stage optimisation problem. Following Enevoldsen and Sørensen (1994), reliability-based optimisation problems can be treated within the framework of pre-posterior decision analysis. If benefits are neglected, the optimisation becomes a minimisation of expected total cost C, or Bayes risk:

$$\min_{e,d} E_{\theta}\left[C_T(\theta|e,d)\right] \qquad (6.12)$$

Typically, the minimisation is constrained on societal requirements for failure probability; $P_{F,max}$, and on physical bounds of the optimisation variables, here written for the case of reliability based structural optimisation (RBSO). The first is a nonlinear constraint, whereas the latter are linear bounds.

$$\min_{z} E_{\theta}\left[C_F(z)P_F(\theta|z)\right]+C_I(z)$$
$$s.t.$$
$$P_{F,i}(z) \leq P_{F,i}^{max}, i=1,\ldots,n \qquad (6.13)$$
$$z_j^l \leq z_j \leq z_j^u, j=1,\ldots,m$$

In the above, C_I represents construction costs, C_F the failure costs, $z \in Z$ are the design variables and z_l and z_u are upper and lower bounds on the design variables. It should be noted that the expected failure costs in equation 6.13 can be neglected if there are requirements on the failure probability.

In the case of inclusion of events like inspections, repairs or experiments (equation 6.14) still apply when the cost function is made to include the respective cost functions. For an SHM system, writing out the corresponding expression, and averaging over $\boldsymbol{\theta}$, we obtain

$$\min_{e,d} C_{\text{ins}}(e,d) + C_R(e,d) + C_F(e,d) + C_1(e)$$

$$s.t. \tag{6.14}$$

$$P_{F,i}(z) \le P_{F,i}^{\max}, i = 1,\dots,n$$

where C_{ins} are the expected inspection costs, C_R are the expected repair costs, C_F are the expected failure costs and C_1 are the expected initial costs.

6.4.2 Optimising the detector

The simplest form of detector is a binary hypothesis test. The detector classifies data into two states – null and alternate – so the detector is a threshold in the range of the features. The detector characteristics can be calculated directly for any choice of threshold if closed-form distributions for the features, $p(x|\theta_j)$, are known. If they are not available, which is typically the case, the statistics can be estimated by sampling in a supervised learning setting. With the feature statistics known, a pre-posterior decision analysis is used to optimise the detector threshold (i.e. the optimal detector is that decision which minimises the expected costs, given the feature data available, when taken).

For any time-variant problem, the decisions are discrete. Thus, an optimisation is facilitated by discretising into time slices. The expected costs are then a sum over J states, K decisions and L time slices:

$$\min_{z,d} \sum_{j,k,l} C^e\left(d_k,\theta_j,z\right) P\left(d_k|\theta_j,z\right) P\left(\theta_j|t_l\right) + C^i(z) \tag{6.15}$$

where C^e is the extrinsic cost function corresponding to the confusion matrix and C^i is the cost function associated with initial and SHM system costs. It should be noted that the posterior of the unknown parameters is taken. The total expected costs depend on the choice of inspection and repair strategy.

The simplest strategy (strategy I) is *detection triggers repair*, making the number of branches in the event tree 2^n (Figure 6.20).

Alternatively, SHM can be used as decision support for inspection planning.

Strategy II, shown in Figure 6.21, has 3^n branches. In the following, strategy I is assumed. The probabilities at each time step are given as marginal probabilities of

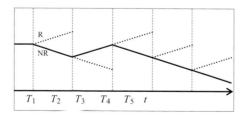

Figure 6.20 Event tree for SHM-based decision support, strategy I. The shown sequence is repair at $t = T_2$.

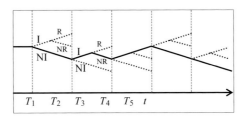

Figure 6.21 Event tree for SHM-based decision support, strategy II. The shown sequence is inspection at $t = T_2$ and inspection + repair at $t = T_3$.

parallel events of N components, given by the event margins defined by each branch of the event tree. The analysis is analogous to inspection updating with inequality information, as described, for example, in Madsen and Sørensen (1990).

$$P_F = P\left(M_1 \leq 0 \cap M_2 \leq 0 \cap ... \cap M_N \leq 0\right) \qquad (6.16)$$

The marginal probability of a parallel system can be sampled with Monte Carlo sampling or estimated to reasonable accuracy with the first-order reliability method (FORM) (Madsen *et al.*, 1986):

$$P_{sys} \approx \Phi\left(-\{\beta\};\{0\},\{\alpha\}^T\{\alpha\}\right) \qquad (6.17)$$

For the event tree in Figure 6.20, defining failure for $M \leq 0$ and detection for $H \leq 0$, for the time up to the second system run we obtain

$$
\begin{aligned}
&\textit{for } 0 \leq t \leq T_1 \\
&\quad P_F(t) = P(M \leq 0) \\
&\textit{for } T_1 \leq t \leq T_2 \\
&\quad P_F(t) = P\left(M(T_1) \leq 0\right) \\
&\qquad + P\left(M(T_1) > 0 \cap H > 0 \cap M^0(t) \leq 0\right) \\
&\qquad + P\left(M(T_1) > 0 \cap H \leq 0 \cap M^1(t) \leq 0\right)
\end{aligned} \qquad (6.18)
$$

where M^0 is the failure margin in case of no detection at $t = T_1$ and M^1 is the failure margin in case of detection at $t = T_1$.

Correspondingly, the probability of repair is

$$
\begin{aligned}
P_R(T_1) &= P\left(M(T)_1 > 0 \cap H \leq 0\right) \\
P_R(T_2) &= P\left(M(T_1) > 0 \cap H > 0 \cap M^0(T_2) > 0 \cap H^0 \leq 0\right) \\
&\quad + P\left(M(T_1) > 0 \cap H \leq 0 \cap M^1(T_2) > 0 \cap H^1 \leq 0\right)
\end{aligned} \qquad (6.19)
$$

The total expected repair costs can be expressed as

$$C_{T,R}\left(N_r,t_r,d\right) = \sum_{i=1}^{N_r} C_R P_{R,i} \qquad (6.20)$$

where N_r is the number of system runs, t is the time between runs, $P_{R,i}$ is the posterior probability of repair at the ith run, given that failure has not occurred earlier, and $C_{R,i}$ is the cost of repair at the ith run.

The total expected failure costs can be expressed as

$$C_{T,F}\left(N_r, t_r, d\right) = \sum_{i=1}^{N_r+1} C_F\left(T_i\right) \cdot \left(P_F\left(T_i\right) - P_F\left(T_{i-1}\right)\right) \tag{6.21}$$

where $C_F(T)$ is the cost of failure at time T and $P_F(T)$ is the posterior failure probability. To account for the real rate of interest r, the sum-terms in expressions 6.20 and 6.21 are multiplied by

$$(1+r)^{-T_i} \tag{6.22}$$

As the reliability evaluation is nested within the optimisation, the optimisation is a double loop. In general, solving RBSO problems require constrained nonlinear optimisation algorithms.

In the case of an isolated structure with no societal requirements for the failure probability, the problem reduces to a risk optimisation.

6.5 Example, continued: designing the SHM system

In the current example we assume that fatigue only occurs in one location, chosen as the tower foot, on the wind turbine tower. This corresponds to designing a localised monitoring system, and makes the problem more manageable for a brief example.

6.5.1 Choice of damage-sensitive feature

To link the deterioration state D of the Palmgren–Miner hypothesis to the secant modulus of the concrete in the damaged region, E_{fat}, the data produced from a stress ratio of 0.6 by Holmen and the uncertainty model from the Breitenbücher and Ibuk (2006) data were interpolated. The damage-sensitive feature is thus selected as $v_1 = E_{fat}$, where E_{fat} is a random function of D, as shown in Figure 6.15. The function is parameterised into

$$E_{fat}/E_{sec,0} = f(D, c)$$

where f() is a deterministic function of D and c, c is a time-independent random variable, fitted to model the variation described in Breitenbücher and Ibuk (2006), and D is a deterministic function of several time-invariant random variables, all given in Table 6.2, and of the loading, which is also considered time-invariant.

6.5.2 Designing the detector

In the present example, the sensor network design is omitted from the scope by using an adaptation of the Bayesian concept of expected value of perfect information (EVPI), first introduced in Raiffa and Schlaifer (1961). In brief, this theoretical value is a measure of the economic benefit the 'perfect experiment' can be expected to yield. The Bayesian perfect experiment is in fact not an experiment, but is certain insight, for the decision-maker, into the otherwise unknown state. It is thus equal to an experiment

Table 6.3 Cost matrix for damage detection.

	State = D_0	State = D_1
Decide d_0	True negative $C_{00} = 0$	False negative $C_{01} = C_F$
Decide d_1	False positive $C_{10} = C_{rep}$	True positive $C_{11} = C_{rep}$

without noise and uncertainty. In this study, adapting the EVPI strictly means that the deterioration state D is observed without uncertainty at every sensing instance. However, an adaptation is used, where, rather than D, which is in fact an unobservable – or hidden – variable, the damage-sensitive feature v_1 is observed without uncertainty. By using this adaptation, it becomes possible to estimate the most value that a SHM system can possibly bring, when designed to target this specific feature. The EVPI is calculated using normal probability analysis, using an attached cost function C. For most applications, the discrete cost function shown in Table 6.3 is adequate.

The cost function only contains two event costs; cost of failure C_F and cost of locating the damage and repairing the structure C_{rep}. An inspection cost different from the repair cost could be included, but studies show that the repair cost has very little influence on the optimum of the decision strategy d and on the total expected costs (Hovgaard, 2015). The strategy of detection triggers perfect repair along with a decision strategy d = static detector threshold d_{th}, and a constant SHM system run interval is used. The optimal Bayesian detector minimises the total expected costs by optimising the decision strategy d:

$$d_{opt} \leftarrow \min_d \left(E[C] \right) \tag{6.23}$$

In this case, the total expected costs are then minimised w.r.t. the detector threshold d_{th}. The total expected costs are a sum of the expected costs of four events and the initial costs C_{ini}:

$$E(C) = E[C_{00}] + E[C_{01}] + E[C_{10}] + E[C_{11}] + C_{ini} \tag{6.24}$$

If the type of SHM system to be used is undecided, the analysis becomes a pre-posterior analysis, and the initial costs C_{ini} must be included in the optimisation. When the system is already implemented, the object function reduces to

$$E[C] = \left(P_{10} + P_{11} \right) C_{rep} + P_{01} C_F \tag{6.25}$$

This constitutes the object function for optimisation, with the optimisation variables being SHM parameters (e.g. number of system runs and following decisions in the service life, and decision threshold th). These parameters are both visualised in Figure 6.22.

In the top plot of Figure 6.22, realisations of Miners damage D are shown. The damage progression is fully described by the straight line between ($n = 0, D = 0$) and

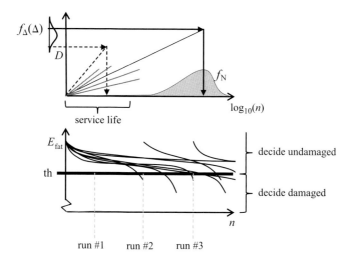

Figure 6.22 Decision-making process for a static decision rule and fatigue SHM. Top: Realisations of the theoretical damage progression. Bottom: Corresponding realisations of the damage-sensitive feature E_{fat} when three SHM instances are performed.

($D = \Delta$) and, due to various epistemic and aleatory uncertainties, the damage rate dD/dn is different for different realisations. In the bottom plot, realisations of the observable variable E_{fat} are sketched. In the sketched example, the SHM system is run three times. At each run, an estimation of E_{fat} is performed and a decision about repair is made based on the result. In offshore engineering, a usual assumption to make is that the repaired structure is fully correlated to the original structure. This means the damage progression is reset to $D = 0$ at time of repair, and that damage rate dD/dn is unchanged.

The optimisation could, in theory, be performed using FORM systems analysis (known from risk-based inspection), as described in Section 6.4.1, using an optimiser capable of constrained nonlinear optimisation with mixed real numbers and integers (e.g. the genetic algorithm). However, such an optimisation quickly becomes intractable when the number of instances becomes high. If the problem is of 300 SHM runs, more than 10^{100} parallel systems need to be analysed by FORM for each iteration of the optimiser. Other approaches such a Bayesian nets (BNs) have been investigated by the author (Hovgaard and Brincker, 2016). The advantage is that the BN structure readily represents the SHM problem, and the effective junction tree algorithm enables efficient inference when the Markovian assumptions are satisfied (e.g. in a hidden Markov model, HMM).

In the present study, a grid search of numerical Monte Carlo simulations of 10^7 samples was used instead, as the exact solution is best approximated by crude Monte Carlo sampling of time histories. The service life was discretised into 10^4 time steps and the number of system runs was chosen to be between 4 and 300. The expected capitalised costs surface and the optimum point for the number of system runs and the static decision threshold are visualised by the surface formed in the grid points. The results are plotted in Figure 6.23.

153

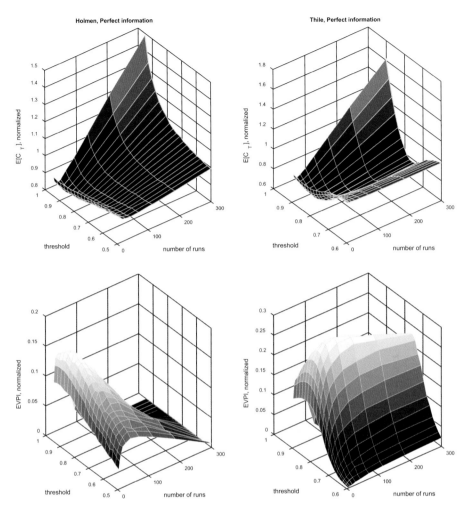

Figure 6.23 Normalised expected capitalised costs given perfect information of the damage-sensitive feature. Left: Using Holmen and Breitenbücher & Ibuk data. Right: Using Thiele data. Each data point is the result of Monte Carlo simulation, using 10^7 sampled time histories.

The optimisation results are listed in Table 6.4.

It can be seen that the performance decreases with a higher number of system runs. The optimal decision threshold for the damage-sensitive feature can be esti-mated from the surfaces created by the grid search. The reduction of normalised total expected costs (= $EVPI$) is seen to be 16% of expected failure costs when using the model from Holmen and Breitenbücher/Ibuk and 29% when using the model from Thiele. This large difference in performance is due to the uncertainty models, which proves the importance of Thiele's findings. It must be recalled, however, that these are optimum values of benefit, obtained given the assumption that the damage-sensitive feature can be observed without uncertainty. In real applications, this assumption does

Table 6.4 Optimisation results.		
Variable	Holmen, Breitenbücher and Ibuk	Thiele
Threshold	0.85	0.79
Runs	27	70
EVPI	0.16	0.29

not hold and measurement noise will contaminate the data, causing the actual VoI to be lower than the *EVPI*. The *EVPI* can then be used to select an SHM monitoring scheme in a pre-posterior fashion, as the implementation and maintenance costs must be lower than the *EVPI* for the system to operate economically.

6.6 Conclusions and further directions

Concrete high-cycle fatigue has been described from an empirical angle, and the framework of cost-effective risk-based inspection strategies has been introduced into the topic of concrete fatigue health monitoring. Design for concrete fatigue has been implemented for decades, but the almost complete absence of failures attributed to fatigue indicates that the design rules are overly conservative. This provides economic grounds for an SHM health monitoring scheme, as overly conservative design rules directly lead to non-economic use of materials.

SHM of concrete fatigue can be performed by monitoring a damage-sensitive feature, that is, an observable parameter that is correlated to the theoretical damage level from the design equations, and that possesses good noise properties. J.O. Holmen was the first to observe a correlation between secant stiffness and theoretical damage, and with the recent results from Thiele, several damage-sensitive features have been investigated, included secant stiffness. In this chapter, the original findings from Holmen have been compared with the model from Thiele in a calculation of expected capitalised costs, and a great difference was observed in performance.

In the design of any structure, the expected costs of failure must be weighed against the cost of construction. In the design of any operations and maintenance strategy, two additional costs are added: cost of repair and cost of inspection. In the case of SHM, the additional costs become costs of false negatives (=failure), false positives (false alarm) and true positives (=repair), and there is also the cost of implementing and maintaining the system. The Bayesian pre-posterior decision analysis framework provides the method of designing and operating an expectedly economic SHM system, but the analysis is not easily carried out without some simplifications. Various approaches were given in this chapter that should enable the designer to calculate the expected value of perfect information *EVPI*, which is the most basic value property of an SHM monitoring strategy, and enables the selection of am SHM system up front.

The *EVPI* is only the first basic analysis in the design of an SHM system. For most structures the damage-sensitive feature E_{fat} cannot be obtained directly using stress–strain measurements, but merely assessed using measurements of correlated variables. This chapter has not dealt with the analysis of such measurements, but background on the subject can be found in Thiele (2015), for example. The full implementation of an SHM system, including vibration (modal) based feature extraction is treated in Hovgaard (2015).

The spatial extent of the fatigue damage 'affected region' has yet to be studied. The primary reason for the absence of knowledge is that all experiments have been carried out on small cylindrical specimens. The actual progression of damage in a large specimen will be effected by the redistribution of stresses in the specimen. As damage affects the stiffness, the stress analysis becomes dependent on the damage distribution, and a non-linear time history analysis must be carried out to estimate the damage progression. Finally, the parameters of a Paris-type power law can be estimated. Such work will form the scope of future research.

References

Aas-Jakobsen, K. (1970) *Fatigue of concrete beams and columns*, PhD thesis, University of Trondheim, Trondheim, Norway.

Basquin, O.H. (1910) The exponential law of endurance tests. In *Proceedings of ASTM* **10**, II.

Bazant, Z.P. and K. Xu (1991) Size effect in fatigue fracture of concrete. *ACI Materials Journal* **88**(4), 390–399.

Bazant, Z.P. and M.H. Hubler (2014) Theory of cyclic creep of concrete based on Paris law for fatigue growth of subcritical microcracks. *Journal of the Mechanics and Physics of Solids* **63**, 187–200.

Bourinet, J.-M., C. Mattrand and V. Dubourg (2009) A review of recent features and improvements added to FERUM software. In *Proceedings of the 10th International Conference on Structural Safety and Reliability (ICOSSAR'09)*, Osaka, Japan.

Breitenbücher, R. and H. Ibuk (2006) Experimentally based investigations on the degradation-process of concrete under cyclic load. *Materials and Structures* **39**, 717–724.

CEB (1993) *CEB-FIP Model Code 90. Bulletin d'Information*, No. 213/214. Thomas Telford, London.

Cornelissen, H. (1984) Fatigue failure of concrete in tension. *HERON* **29**(4), 68.

Cornelissen, H. and H. Reinhardt (1982) Fatigue of plain concrete in uniaxial tension and in alternating tension–compression loading. In *IABSE Colloquium on Fatigue of Steel and Concrete Structures*, Lausanne, Switzerland, 273–282.

Der Kiureghian, A., T. Haukaas and K. Fujimura (2006) Structural reliability software at the University of California, Berkeley. *Structural Safety* **28**(1–2), 44–67.

DNV (2012) *Offshore concrete structures*. DNV, Oslo.

Enevoldsen, I. and J. Sørensen (1994) Reliability-based optimization in structural engineering. *Structural Safety* **15**(3), 169–196.

European Committee for Standardization (2004) *Eurocode 2: Design of concrete structures – Part 1-1: General rules and rules for buildings*, EN 1992-1-1. CEN, Brussels.

Farrar, C.R. and K. Worden (2012) *Structural health monitoring: a machine learning perspective*. Wiley, New York.

Feret, R. (1906) *Etude experimentale du ciment arme*, Chapter 3. Grauthier–Villiers, Paris (in French).

fib (2012) *Model code 2010, final draft*, Volumes 1 and 2. International Federation for Structural Concrete, London.

Flynn, E.B. (2010) *A Bayesian experimental design approach to structural health monitoring with application to ultrasonic guides waves*. Doctoral thesis, UC San Diego, San Diego, CA.

Gaspar, R. (2012) *Concrete wind towers – a low-tech innovation for a high tech sector*, thisisxy. Available at http://www.thisisxy.com/blog/concrete-wind-towers-a-low-tech-innovation -for-a-high-tech-sector (accessed 17 August 2016).

Grünberg, J. and J. Göhlmann (2006) Schädigungsberechnung an einem Spannbetonschaft für eine Windenergieanlage unter mehrstufiger Ermüdung. *Beton und Stahlbetonbau* **101**(8), 557–570.

Hillerborg, A., M. Modéer and P.-E. Petersson (1976) Analysis of crack formation and crack growth in concrete by means of fracture mechanics and finite elements. *Cement and Concrete Research* **6**(6), 773–781.

Holmen, J. (1979) Fatigue of concrete by constant and variable amplitude loading. Division of Concrete Structures, Norwegian Institute of Technology, University of Trondheim, Trondheim.

Holmen, J. (1984) Fatigue design evaluation of offshore concrete structures. *Matériaux et Construction* **17**(1), 39–42.

Hovgaard, M.K. (2015) *Incorporating structural health monitoring in the design of slip formed concrete wind turbine towers*. PhD thesis, River Publishers, Aalborg.

Hovgaard, M.K. and R. Brincker (2016) Limited memory influence diagrams for structural damage detection decision-making. *Journal of Civil Structural Health Monitoring* **6**(2), 205–215.

IEC (2005) *Wind turbines – Part 1: Design requirements*, IEC 61400-1:2005. International Electrotechnical Commission, Geneva.

JCSS, Joint Committee on Structural Safety (2006) *Probabilistic model code*. JCSS, Zurich. Available online at http://www.jcss.byg.dtu.dk/Publications/Probabilistic_Model_Code.aspx.

Jonkman, J., S. Butterfield, W. Musial and G. Scott (2009) Definition of a 5-MW reference wind turbine for offshore system development. National Renewable Energy Laboratory, Colorado.

Lassen, T. (1997) *Experimental investigation and stochastic modelling of the fatigue behaviour of welded steel joints*. River Publishers, Aalborg.

Lassen, T. and N. Recho (2006) *Fatigue life analyses of welded structures*. Iste, London.

Lindley, D. (1971) *Bayesian statistics, a review*. Society for Industrial and Applied Mathematics, Philadelphia, PA.

Lohaus, L., N. Oneschkow and M. Wefer (2021) Design model for the fatigue behaviour of normal-strength, high-strength and ultra-high-strength concrete. *Structural Concrete* **13**, 182–192.

Madsen, H., S. Krenk and N. Lind. (1986) *Methods of structural safety*. Prentice Hall, Englewood Cliffs, NJ.

Madsen, H. and J. Sørensen (1990) Probability-based optimization of fatigue design, inspection and maintenance. In *Proceedings from the Fourth Symposium on Integrity of Offshore Structures*, Glasgow, pp. 327–334.

McCall, J.T. (1958) Probability of fatigue failure of plain concrete. *ACI Journal Proceedings* **55**(8), 233–244.

Miner, M.A. (1945) Cumulative damage in fatigue. *Journal of Applied Mechanics* **12**(67), A159–A164.

Ocean Structures (2009) *Ageing of offshore concrete structures*, OSL-804-R04, Rev. 2. Ocean Structures, Myreside Steading.

Oh, B. (1986) Fatigue analysis of plain concrete in flexure. *Journal of Structural Engineering* **112**(2), 273–288.

Palmgren, A. (1924) Die Lebensdauer von Kugellagern. *Zeitschrift des Vereins Deutscher Ingenieure* **68**(14), 339–341.

Paris, P.C. and F. Erdogan (1963) A critical analysis of crack propagation laws. *Journal of Fluids Engineering* **85**(4), 528–533.

Petković, G., R. Lenschow, H. Stemland and S. Rosseland (1990) *Fatigue of high-strength concrete*, ACI Special Publication 121. ACI, Detroit, MI, pp. 505–525.

Petryna, Y. and W. Krätzig (2005) Computational framework for long-term reliability analysis of RC structures. *Computer Methods in Applied Mechanics and Engineering* **194**(12), 1619–1639.

Raiffa, H. and R. Schlaifer (1961) *Applied statistical decision theory*. Division of Research, Harvard Business School, Boston, MA.

RILEM Committee 36-RDL (1984) Long term random dynamic loading of concrete structures. *Materiaux et Constructions* **17**, 97.

Shah, S.P. (1982) *Fatigue of concrete structures*. American Concrete Institute, Farmington Hills, MI.

Sobol, J.-M. (2001) Global sensitivity indices for nonlinear mathematical models and their Monte Carlo estimates. *Mathematics and Computers in Simulation* **55**, 271–280.

Sørensen, J. and H. Toft (2010) Probabilistic design of wind turbines. *Energies* **3**(2), 241–257.

Stemland, H., G. Petkovic, S. Rosseland and R. Lenschow (1990) Fatigue of high strength concrete. *Nordic Concrete Research* **90**, 172–196.

Straub, D. (2004) *Generic approaches to risk based inspection planning for steel structures*. Institute of Structural Engineering, ETH, Zurich.

Tepfers, R. (1979) Tensile fatigue strength of plain concrete. *ACI Journal* **76**(8), 919–933.

Tepfers, R. and T. Kutti (1979) Fatigue strength of plain and ordinary and lightweight concrete. *ACI Journal* **76**(5), 635–652.

Thiele, M. (2015) *Experimentelle Untersuchung und Analyse der Schädigungsevolution in Beton unter hochzyklischen Ermüdungsbeanspruchungen*. Thesis, Technical University of Berlin, Berlin.

Thoft-Christensen, P. and Baker, M.J. (1982) *Structural reliability theory and its applications*. Springer Verlag, Berlin and Heidelberg.

Urban, S., A. Strauss, R. Schütz, K. Bergmeister and C. Dehlinger (2014) Dynamically loaded concrete structures – monitoring-based assessment of the real degree of fatigue deterioration. *Structural Concrete* **15**, 530–542.

Wefer, M. (2010) *Materialverhalten und Bemessungswerte vonultrahochfestem Beton unter einaxialer Ermüdungsbeanspruchung*. Dissertation, Leibniz University of Hannover, Institute of Building Materials Science, Leibniz.

Wöhler, A. (1860) Versuche über die Festigkeit der Eisenbahnwagenachsen. *Zeitschrift für Bauwesen* **10**, 160–161.

Appendix A
MATLAB code for probabilistic model of concrete fatigue

```matlab
function [N, cas] = cyclesToFailure(m, s, fc, fct, XNc, XNt, beta_cc)
% calculates the number of cycles to failure for a given
%    m  - mean level (N/m^2)
%    s  - stress range (N(m^2)
%    fc - cylinder strenght (N/m^2)
%
% definitions: tension is positive (m)
%
% based on FIB MC2010 - final draft - 2012 - vol 1, Method II verification.
% for compression AND cornelissen (1984)
%
% copyright. Mads Hovgaard. Rambøll. January 2014.
% v3: vectorized
% ▫▫▫▫▫▫▫▫▫▫▫▫▫▫▫▫▫▫▫▫▫▫▫▫▫▫▫▫▫▫▫▫▫▫▫▫▫▫▫▫▫▫▫▫▫▫▫▫▫▫▫▫▫▫▫▫▫▫▫▫▫▫▫▫▫▫▫▫▫▫▫▫▫
if nargin < 7
    beta_cc = 1;
end
%%
m           = m(:);                              % force col vec
s           = s(:)';                             % force row vec
M           = repmat(m,1,length(s));
S           = repmat(s,length(m),1);
smin        = M - S./2;
smax        = M + S./2; % smax > smin, generally!
beta_c_sus  = 0.85;
fcfat       = beta_cc*beta_c_sus*fc*(1-(fc)/400);      % (5.1-110)
cas1        = zeros(size(M));
cas2        = zeros(size(M));
cas3        = zeros(size(M));
cas1(smax <= 0) = 1; % compression
cas2(smax > 0 & smin <= 0) = 1; % alternating tension and smin = 0
cas3(smin >= 0) = 1; % pure tension with smin >= 0

%% case 1 - compression
scmin1      = abs(smax);                       % minimum compressive stress
scmax1      = abs(smin);                       % maximum compressive stress
nu          = 1./(1.5 - 0.5*scmin1./scmax1);          % (7.4-2)
nu(cas1==0) = nan;
Sc_min1     = scmin1.*nu./fcfat;
Sc_max1     = scmax1.*nu./fcfat; % maximum stress ratio ( <= 1)
Sc_min1(Sc_min1>0.8) = 0.8;
Y           = (0.45+1.8.*Sc_min1)./(1+1.8.*Sc_min1-0.3.*Sc_min1.^2);
tmp1        = XNc*(8./(Y-1).*(Sc_max1-1));
tmp1(tmp1>XNc*8) = 0;                          % (5.1.109a)
tmp2        = XNc*(8 + (8*log(10))./(Y-1) .* (Y-Sc_min1) .* log10((Sc_max1-Sc_min1)./(Y-Sc_min1))); % (5.1.108)
```

159

```
tmp2(tmp2<=XNc*8) = 0;                                    % (5.1.109b)
logN1          = tmp1+tmp2;                                % (5.1.107)
logN1(Sc_max1>1) = 0;

%% case 2 - alternating
% tensile failure
smax21         = smax;                                    % tensile stress
smin21         = abs(smin);                                % compressive stress
S_max21        = smax21./fct;
Sc_min21       = smin21./fc;
logN21         = XNt*(9.36 - 7.93.*S_max21 - 2.59.*Sc_min21);   % (6) p.23
Cornellisen (1984)
logN21(S_max21>1) = 0;
% compressive failure - if the compressive strenght for tensile stress = 0
% is less than the tensile failure capacity
scmax22        = abs(smin);                                % maximum
compressive stress
nu             = 1/1.5*ones(size(scmax22));               % (7.4-2)
nu(cas2==0) = nan;
Sc_max22       = scmax22.*nu./fcfat;
Y              = (0.45+1.8.*0)./(1+1.8.*0-0.3.*0.^2);
tmp1           = XNc*(8./(Y-1).*(Sc_max22-1));
tmp1(tmp1>XNc*8) = 0;                                      % (5.1.109a)
tmp2           = XNc*(8 + (8*log(10))./(Y-1) .* (Y-0) .* log10((Sc_max22-0)./
(Y-0)));  % (5.1.108)
tmp2(tmp2<=XNc*8) = 0;                                     % (5.1.109b)
logN22         = tmp1+tmp2;                                % (5.1.107)
logN2          = logN21;
tmp            = logN21>logN22;
logN2(tmp)     = logN22(tmp);

%% case 3 - pure tension with smin >= 0
smax3          = smax;                                    % highest tensile stress
smin3          = smin;                                    % lowest tensile stress
S_max3         = smax3./fct;
S_min3         = smin3./fct;
logN3          = XNt*(14.81 - 14.52.*S_max3 - 2.79.*S_min3);   % (5) p.23
Cornellisen (1984)
logN3(S_max3>1) = 0;

%% collect
logN1(cas1==0) = 0;
logN2(cas2==0) = 0;
logN3(cas3==0) = 0;
logN           = logN1 + logN2 + logN3;
% logN = logN1;
N              = 10.^logN;
cas            = cas1;
cas(cas2==1) = 2;
cas(cas3==1) = 3;
```

Chapter 7
Validation of models for prediction of chloride ingress in concrete exposed to a de-icing salt road environment

Luping Tang

7.1 Introduction

At the present time, the specification of durability is mainly based on the establishment of various constraints on the mixture proportions of concrete, such as cement type and content, water–binder ratio, entrained air content or cover thickness, as a function of severity of exposure. This approach does not consider the actual performance of concrete materials that have different types of cement and mineral components added to the cement or directly to the concrete. Regarding the sustainable development of society and the construction industry, there is a tendency to use performance-based service life design and calculations instead of prescription-based approaches. With the help of increasingly sophisticated durability models, safer structures can be designed with an expected service life and a reduction in the consumption of materials. In recent years a number of models for service life design have been suggested by relevant national, regional or international committees, mainly dealing with durability problems, especially chloride-induced reinforcement corrosion. In Sweden, with its long coastline and cold climate and thus intensive use of de-icing salt, great efforts have been made in the development of models for chloride ingress into concrete. To validate the models for the de-icing salt road environment, efforts have also been made to collect data from structures exposed to weathering. This has been necessary in order to be able to use the models in service life design or redesign of reinforced concrete structures. This chapter will present models for the prediction of chloride ingress and their validation against data from exposed structures.

7.2 Models for prediction of chloride ingress

Due to the global problems of chloride-induced corrosion of reinforcement in concrete structures, the topic of chloride ingress in concrete has been intensively studied in recent decades. An overview of the durability of steel-reinforced concrete in chloride environments was given by Shi *et al.* (2012), in which different test methods for chloride diffusivity and challenges in assessing the durability of concrete from its chloride diffusivity were also discussed.

With regard to models of chloride ingress, many models can be found in the literature. Some typical models have been summarised by Tang *et al.* (2011).

According to them, the models can be categorised into two groups: empirical and mechanistic. Empirical or semi-empirical models often assume a diffusion process but use total chloride content as the driving force. Obviously, taking the total chloride content as a driving force as the transference function is theoretically questionable, because it is only the free chloride ions that can move in the pore solution and contribute to chloride-induced corrosion of reinforcement in concrete. Therefore, mechanistic models often use the free chloride as the driving force and take non-linear chloride binding into account. On the other hand, for the de-icing road environment or splash zone of the marine environment, chloride transport in concrete involves both capillary suction and diffusion. It is believed that the rate of capillary suction should be larger than that of diffusion. Due to the effect of chloride binding, the chloride ingress profiles under the condition of wetting–drying (capillary suction, diffusion and evaporation/deposition) are, however, apparently similar to those obtained under the saturated condition (pure diffusion), as numerically simulated by Meijers *et al.* (2005) and experimentally demonstrated by Gang *et al.* (2015). Therefore, models based on the diffusion process also seem to be suitable for chloride ingress related to de-icing road environments. Other models listed in the literature are briefly presented as follows.

7.2.1 The simple ERFC model

The abbreviation ERFC denotes the mathematical symbol erfc (ERror Function Complement). This model was first proposed in the early 1970s for the modelling of chloride ingress into concrete (Collepardi *et al.*, 1972). The model uses an erfc solution to Fick's 2nd law of diffusion under the semi-infinite boundary condition:

$$C(x,t) = C_i + (C_s - C_i) \cdot \mathrm{erfc}\left(\frac{x}{2 \cdot \sqrt{D_a \cdot t}} \right) \qquad (7.1)$$

where C_i is the initial chloride content in the concrete (sometimes this chloride content is negligible), C_s is the surface chloride content, x is depth, D_a is the apparent diffusion coefficient and t is exposure duration. In this model, parameters C_s and D_a are assumed constant during the whole period of exposure.

In this model the key parameters are C_s and D_a, which have to be determined by using curve-fitting of the chloride ingress profiles from on-site or laboratory exposure. It has been proven from considerable experimental data that this simple model can only describe the chloride ingress under the exposure conditions for a short duration close to the conditions for which the input parameters were determined. As explained later, several modifications to equation 7.1 have been proposed to try to expand the applicability range of such types of model.

7.2.2 Mejlbro–Poulsen's Model

This mathematical model was developed in Denmark by the Danish national project HETEK in the mid-1990s (Frederiksen *et al.*, 1996). The model assumes that total chloride is the driving force, and considers both the surface chloride content C_s and the apparent diffusion coefficient D_a as time-dependent functions, that is:

$$D_a = D_{aex} \left(\frac{t_{ex}}{t + t_{ex}} \right)^\alpha \qquad (7.2)$$

and

$$C_s = C_i + S \left(D_a \cdot t \right)^p \qquad (7.3)$$

where D_{aex} is the apparent diffusion coefficient at the time of exposure t_{ex}, and α, S and p are constants. An analytic solution to Fick's 2nd law with time-dependent C_s and D_a was given in the form as follows (Mejlbro, 1996):

$$C(x,t) = C_i + \left(C_s - C_i \right) \cdot \Psi_p \left(\frac{x}{2\sqrt{D_a \cdot t}} \right) \qquad (7.4)$$

where Ψ is a series of Γ functions (gamma-functions). When $p = 0$, equation 7.4 becomes the same form as equation 7.1. Because it is impossible to measure D_{aex} at the time of exposure (no ingress profile is available for curve-fitting), the model needs some experimental data from short-term exposures (e.g. one year exposure and at a later storage) to estimate the values of D_{aex}, α, S and p. Updating from later available site data may result in different values for these empirical parameters (Frederiksen and Geiker 2000, 2008).

7.2.3 DuraCrete Model

The DuraCrete project (Engelund *et al.*, 2000) recommended the following equation to express the apparent diffusion coefficient in equation 7.1:

$$D_a = k_{e,cl} \cdot k_{c,cl} \cdot D_{RCM,0} \cdot \left(\frac{t_0}{t + t_{ex}} \right)^{n_{cl}} \qquad (7.5)$$

where $D_{RCM,0}$ is the chloride migration coefficient measured, for example, by the Nordtest method NT BUILD 492 (Nordtest, 1999) at the age of $t_0 = 28$ days, $k_{e,cl}$ and $k_{c,cl}$ are constants that consider the influence of environment and curing, respectively, on chloride ingress, t_0 is the reference period (concrete age of 28 days) at which $D_{RCM,0}$ is measured and n_{cl} is the age factor describing the time dependency of the apparent diffusion coefficient.

It can be seen that there is a difference between equations 7.2 and 7.5 in that D_{aex} in equation 7.2 is from the same site exposure conditions as D_a, while $D_{RCM,0}$ in equation 7.5 is from the laboratory conditions, which are different from the actual exposure conditions. The model uses an empirical factor $k_{e,cl}$ to try to bridge the gap between the laboratory and on-site conditions. Again, it requires a lot of qualified on-site data to establish the proper values of $k_{e,cl}$ for the actual service life design. The principle of the DuraCrete model was also adopted in guidance from the Concrete Society (Bamforth,

2004) and the fib model code for service life design (fib, 2006), in which factor $k_{e,cl}$ was specified by

$$k_{e,cl} = \exp\left[b_e \left(\frac{1}{273+T_{ref}} - \frac{1}{273+T_{real}} \right) \right] \qquad (7.6)$$

where b_e is the regression variable, which varies between 3,500 and 5,500, with 4,800 used as the mean value and 700 as the standard deviation, T_{ref} is the reference temperature in °C at which the chloride migration coefficient is measured, and T_{real} is the real exposure temperature in °C.

7.2.4 ACI Life 365 model

In North America, an empirical model was developed by ACI TC-365 in the early 2000s (Thomas and Bentz, 2001). The model utilises Fick's 2nd law of time-dependent diffusion as the transport function with the total chloride content C as the driving force:

$$\frac{\partial C}{\partial t} = D(t)\frac{\partial^2 C}{\partial x^2} \qquad (7.7)$$

and

$$D(t) = D_{ref}\left(\frac{t_{ref}}{t}\right)^m \qquad (7.8)$$

where D_{ref} is the apparent diffusion coefficient at the reference time of exposure t_{ref} and m is a constant. In order to prevent the diffusion coefficient indefinitely decreasing with time, the relationship shown in equation 7.7 is only valid up to 30 years. Beyond that time, the value at 30 years (D_{30y}) calculated from equation 7.7 is assumed to be constant throughout the rest of the analysis period. The temperature effect on the apparent diffusion coefficient has also been taken into account in this model. The model selects the rate of chloride build-up and the maximum surface content based on the type of exposure, the structure and the geographic location. The model also provides various α values for different additions of pozzolanic materials. Clearly, the model is semi-empiric. On the other hand, the software of the ACI model has integrated chloride ingress, the initiation and propagation of corrosion, the repair schedule and life-cycle costs together, which gives the user a simple tool for maintenance planning of concrete structures. A numerical approach is followed in the model for the time integration. Therefore, special software is needed when using the model.

An important difference that might confuse readers/users of the models is that $D(t)$ in equation 7.8 is different from D_a in equations 7.2 and 7.5. The former is the instantaneous diffusion coefficient as conventionally defined in Fick's law, while the latter is the average diffusion coefficient during the period from t_{ex} to $(t + t_{ex})$ (Poulsen, 1993).

7.2.5 ClinConc Model

The ClinConc model (Cl in Concrete) was first developed in the mid-1990s (Tang and Nilsson, 1994; Tang, 1996). The ClinConc model consists of two main procedures: 1) simulation of free chloride penetration through the pore solution in concrete using a genuine flux equation based on the principle of Fick's law with the free chloride concentration as the driving potential and 2) calculation of the distribution of the total chloride content in concrete using the mass balance equation combined with non-linear chloride binding. Clearly, the ClinConc model uses free chloride as the driving force and takes non-linear chloride binding into account. Thus it describes chloride transport in concrete in a more scientific way than empirical or semi-empirical models. Later, this model was expressed in a more engineer-friendly way (Tang, 2006, 2008) to make it possible to be used by practising engineers. Free chloride concentration in the concrete at depth x is determined using the following equation:

$$\frac{c-c_i}{c_s-c_i} = 1 - \mathrm{erf}\left(\frac{x}{2\sqrt{\frac{\xi_D D_{6m}}{1-n} \cdot \left(\frac{t_{6m}}{t}\right)^n \cdot \left[\left(1+\frac{t_{ex}}{t}\right)^{1-n} - \left(\frac{t_{ex}}{t}\right)^{1-n}\right] \cdot t}}\right) \quad (7.9)$$

where c, c_s and c_i are the concentrations of free chlorides in the pore solution at depth x, at the surface of the concrete and initially in the concrete, respectively, D_{6m} is the diffusion coefficient measured by the rapid chloride migration (RCM) test, e.g. NT BUILD 492, at t_{6m}, ξ_D is the factor bridging the laboratory-measured D_{6m} to the initial apparent diffusion coefficient for the actual exposure environment, n is the age factor accounting for the diffusivity decrease with age, t_{ex} is the age of the concrete at the start of exposure and t is the duration of the exposure.

Different from the empirical models, factors ξ_D and n in ClinConc can be calculated based on the physical properties of the concrete, including cement hydration, hydroxide content, water-accessible porosity, time-dependent chloride binding and environmental parameters such as chloride concentration and temperature. Detailed descriptions of factors ξ_D and n are given in Tang (2006).

The total chloride content C is basically the sum of the bound chloride, c_b, and free chloride, c, expressed as

$$C = \frac{\varepsilon \cdot (c_b + c)}{B_c} \times 100 \text{ mass\% of binder} \quad (7.10)$$

where ε is the water-accessible porosity at the age after exposure, B_c is the cementitious binder content, in kg/m^3 of concrete, and c_b is the bound chlorides expressed as the same unit as free chloride. The bound chlorides can be calculated by different equations, such as Langmuir (Sergi et al., 1992), Freundlich (Tang and Nilsson, 1993) or other regression equations. Detailed descriptions of the use of the Freundlich equation for calculation of bound chlorides are given in Tang (2006).

It should be noticed that the ClinConc model was primarily developed for the submerged environment, where the chloride solution is constantly in contact with the concrete surface. When the model is used for the atmospheric zone or the road environment, certain modifications are needed.

Both the DuraCrete and ClinConc models use the diffusion coefficient measured by the RCM test (e.g. NT BUILD 492) as an input parameter, but care must be taken as this parameter is tested at different concrete ages (i.e. 28 days in the former and 6 months in the latter). Among the above-mentioned models, only the ClinConc model treats the material properties and exposure environment in a separate way.

7.2.6 Other models

As well as the above-mentioned models, there are many other models based on either empirical equations or physical and chemical/electrochemical processes. Some are more or less similar to the Life-365 or DuraCrete models, while some need special software, such as STADIUM®, to calculate the complicated mathematical iterations (e.g. Samson *et al.*, 1999; Truc *et al.*, 2000; Johannesson, 2000; Meijers, 2003; Petre-Lazar *et al.*, 2003; Nguyen *et al.*, 2006; Conciatori *et al.* 2010). Additional information about various models can be found in Tang *et al.* (2011). There are also various numerical models, such as the Multi-Environmental Time Similarity (METS) model (Jin and Xu, 2010) and the LIFEPROD model (Andrade *et al.*, 2014), and so on.

There are also some computational models coupling both material deterioration and thermodynamic/mechanical equilibrium (e.g. Maekawa *et al.*, 2003; Hosokawa *et al.*, 2008).

7.3 Uncertainty in the modelling of chloride ingress

Because the basic outcomes from modelling are chloride concentrations c or contents C, the uncertainty in the output means the quantitative influence of each input parameter has an influence on the modelled output values of c or C. The sensitivities of various input parameters in the error-function-based models have previously been analysed by Tang *et al.* (2011). Based on their analysis, the sensitivity of surface chloride concentration c_s or content C_s is constantly equal to 1, that is,

$$\frac{\Delta c}{\Delta c_s} \cdot \frac{c_s}{c} = \frac{\Delta C}{\Delta C_s} \cdot \frac{C_s}{C} = 1 \tag{7.11}$$

This means that any relative change in surface concentration c_s or content C_s will result in an equal relative change in concentration c or content C. The sensitivities of other parameters are dependent on the ratio of c/c_s or C/C_s (Tang *et al.*, 2011). As an example, for the ClinConc model the sensitivity of D_{6m} or ξ_D can be expressed as

$$\frac{\Delta c}{\Delta D_{6m}} \cdot \frac{D_{6m}}{c} = \frac{\Delta c}{\Delta \xi_D} \cdot \frac{\xi_D}{c} = \frac{1}{\sqrt{\pi}} \cdot \frac{c_s}{c} \cdot z \cdot e^{-z^2} = \frac{1}{\sqrt{\pi}} \cdot \frac{z \cdot e^{-z^2}}{\dfrac{c}{c_s}} \tag{7.12}$$

where

$$z = \frac{x}{2\sqrt{\dfrac{\xi_D D_{6m}}{1-n}\left(\dfrac{t_{6m}}{t}\right)^n \left[\left(1+\dfrac{t_{ex}}{t}\right)^{1-n} - \left(\dfrac{t_{ex}}{t}\right)^{1-n}\right] \cdot t}} \tag{7.13}$$

It is apparent that the sensitivity of D_{6m} and ξ_D is dependent on both c/c_s and z, and the latter contains all the variations except for c_s. Their relationships are illustrated in Figure 7.1. The sensitivity of n can be expressed as

$$\frac{\Delta c}{\Delta D_{6m}} \cdot \frac{D_{6m}}{c}$$

or

$$\frac{\Delta c}{\Delta \xi_D} \cdot \frac{\xi_D}{c}$$

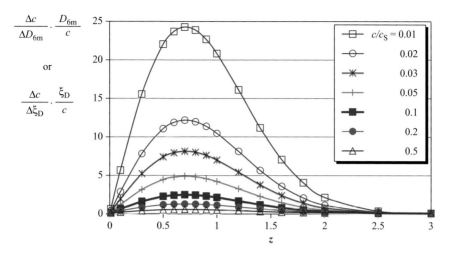

Figure 7.1 Sensitivity of parameters D_{6m} and ξ_D in the ClinConc model.

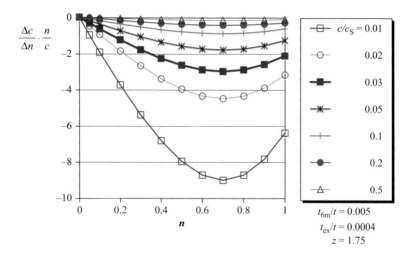

Figure 7.2 Sensitivity of parameter n in the ClinConc model.

$$\frac{\Delta c}{\Delta n} \cdot \frac{n}{c} = \frac{n}{\sqrt{\pi}} \cdot \frac{z \cdot e^{-z^2}}{\dfrac{c}{c_s}} \left[\ln\left(\frac{t_{6m}}{t}\right) + \frac{1}{1-n} - \frac{\ln\left(1+\dfrac{t_{ex}}{t}\right)\cdot\left(1+\dfrac{t_{ex}}{t}\right)^{1-n} - \ln\left(\dfrac{t_{ex}}{t}\right)\cdot\left(\dfrac{t_{ex}}{t}\right)^{1-n}}{\left(1+\dfrac{t_{ex}}{t}\right)^{1-n} - \left(\dfrac{t_{ex}}{t}\right)^{1-n}} \right] \quad (7.14)$$

It can be seen in equations 7.12 and 7.14 that the sensitivity of these input parameters is inversely proportional to the ratio c/c_s. Our important concentration is the threshold concentration for corrosion initiation. When $C_{cr} = 0.4\%$ (by mass of binder) is taken as the criterion, the value of c_{cr} varies between 0.1 and 0.5 g/l, depending on the chloride binding capacity of the binder in the concrete and the initial alkalinity of concrete. In the ClinConc modelling the value of n is normally less than 0.2, with $t_{6m} = 0.5$ years, $t_{ex} = 0.04$ years (two weeks), $t = 100$ years, giving $t_{6m}/t = 0.005$ and $t_{ex}/t = 0.0004$. Assuming a cover thickness of 50 mm, the value of z varies between 1.5 and 2. For $z = 1.75$, the influences of n and c/c_s on the sensitivity of n are illustrated in Figure 7.2.

It can be seen from Figures 7.1 and 7.2 that the sensitivity of the input parameters is strongly dependent on the ratio c/c_s, or more meaningfully c_{cr}/c_s, in relation to corrosion initiation. The influence of parameters D_{6m} and ξ_D on chloride concentration differs from that of parameter n. Any increase in the former leads to an increase in concentration, while any increase in the latter leads to a decrease in concentration. Depending on the individual type of concrete and the assumption of the chloride threshold, the ratio c_{cr}/c_s varies. As a consequence, the sensitivity of the input

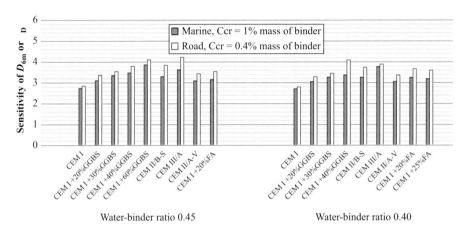

Figure 7.3 Sensitivity of parameter D_{6m} and ξ_D for various types of concrete (Marine = Swedish west coast).

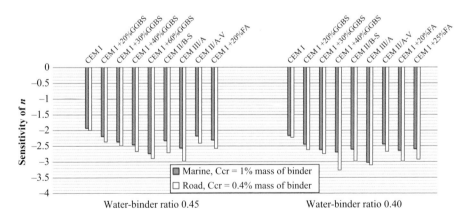

Figure 7.4 Sensitivity of parameter n for various types of concrete (Marine = Swedish west coast).

parameters also varies, as shown in Figures 7.3 (for parameters D_{6m} and ξ_D) and 7.4 (for parameter n). It can be seen that the sensitivities of input parameters for concrete with mineral additions such as ground granulated blast-furnace slag (GGBS) and fly ash (FA) are slightly higher than those with plain Portland cement, because of the higher chloride binding capacity and lower alkalinity of concrete with mineral additions, both of which lead to a lower free chloride concentration, or lower value of c/c_s.

7.4 Validation of models against long-term site data

Three models (described in Section 7.2), the simplest ERFC model, the DuraCrete model and the ClinConc model, were evaluated using the collected long-term on-site data. The other models could not be evaluated due to difficulties with the availability of software or lack of input parameters.

7.4.1 Data for over 10 years of exposure in a road environment

Since 1996, a large number of reinforced concrete specimens with different qualities have been exposed at a site by Highway 40 between Borås and Gothenburg in the western part of Sweden, where de-icing salt was intensively used on the road due to the severe winter climate, as shown in Figure 7.5. Chloride profiles in some of the concrete specimens were measured after 1, 2, 5 and 10 years of exposure. These site data are valuable for the validation of prediction models. A detailed description of the site exposure and concrete specimens has been published elsewhere (Utgenannt, 2004; Tang and Utgenannt, 2007).

Figure 7.5 Exposure site by Highway 40.

Table 7.1 Input parameters used in the DuraCrete model.		
Binder type	100% SRPC	95% SRPC + 5% SF
Curing factor, k_c	0.79	
Environmental factor, k_e	0.68	
Mass% of binder, $A_{s,cl}$	2.57	3.23
Age factor, n	0.65	0.79

Two types of concrete with a water–binder ratio $w/b = 0.4$, one ordinary sulfate-resistant Portland cement (SRPC) concrete and another SRPC+5% SF (silica fume), were used for validation. According to the laboratory test results reported by Tang (2003), the value of D_{RCM6m} for the SRPC concrete with $w/b = 0.4$ was 12.2×10^{-12} m^2/s and for SRPC+5% SF with $w/b = 0.4$ was 4.43×10^{-12} m^2/s. The values of D_{RCM6m} were used for all the models, although in DureCrete (2000) the concrete age t_0 was specified as 28 days. If the same n value was assumed for the period from 28 days to 6 months, there should, however, be no difference in the D_a values calculated using $D_{RCM28d} \cdot (t_{28d})^n$ and $D_{RCM6m} \cdot (t_{6m})^n$.

In the guidelines by DuraCrete (Engelund et al., 2000), there were no values for input parameters directly available for a de-icing salt road environment. Although some values were suggested by the Swedish Concrete Association (Betongföreningen, 2007), it has been demonstrated by Tang and Utgenannt (2007) that these values resulted in a significant underestimation of chloride ingress. The values for the atmospheric zone were therefore used in this evaluation, as listed in Table 7.1. An initial chloride content of $C_i = 0.02\%$ of binder was assumed in the modelling. The same C_s value in the DuraCrete model (2000) was used in the simple ERFC model.

The environmental data used in the ClinConc model included the free chloride concentration $c_s = 1.5$ g/l and the annual average temperature $T = 10°C$ (excluding the freezing period). The initial free chloride concentration in the pore solution was assumed to be 0.002 g/l. The concrete age at the start of exposure was $t_{ex} = 28$ days. Other parameters were similar to those used in the previous studies (Tang, 2006).

The predicted and measured chloride profiles are shown in Figures 7.6 and 7.7, where 'H' denotes the site data measured from the horizontal exposure surface and 'V' denotes the site data measured from the vertical exposure surface. It can be seen that none of the models could predict the chloride ingress precisely, especially when the 5-year profiles were higher than the 10-year profiles for the silica fume concrete. The simple ERFC model significantly overestimated the chloride ingress in both types of concrete. The DuraCrete model overestimated the chloride ingress in the SRPC concrete but underestimated is in the silica fume concrete. In addition, the predicted profiles from 1.5 to 10 years were very close to each other, not really reflecting reality. The ClinConc model revealed relatively better predictions, although the 5-year profile for the silica fume concrete was far from the measured values. Measurement uncertainty in the site data could also be a reason for this unexpected result.

7.4.2 Data from over 30 years exposure of real structures

At the end of the 1990s, a number of road bridges (aged 25 to 30 years) around Gothenburg in Sweden were sampled for chloride ingress profiles (Lindvall, 2001).

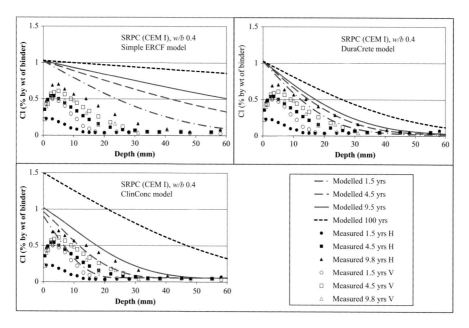

Figure 7.6 Modelled and measured chloride profiles for SRPC concrete.

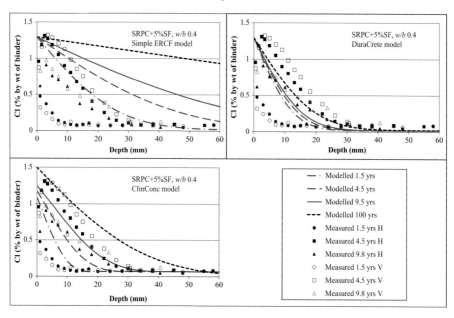

Figure 7.7 Modelled and measured chloride profiles for SRPC+5% SF concrete.

Detailed information about bridges and samplings was published (Lindvall, 2001). Some of the profiles taken from the concrete elements near to the heavy traffic lane were used for validation of the models. The concrete used in the bridges at that time was produced with Swedish SRPC. According to measurements using the RCM test on the specimens taken from the deeper parts of the cores, the D_{RCM} values were in the range of 8.6×10^{-12} to 16.6×10^{-12} m²/s (Lindvall, 2001). Given the age of the concrete (25–30 years), it could be assumed that the water–binder ratio w/c was in the range of 0.4–0.5. Therefore, the mixes with SRPC and $w/c = 0.4$ and 0.5, respectively, were used in the modelling with $D_{RCM6m} = 8.6 \times 10^{-12}$ m²/s for $w/c = 0.4$ and 16.6×10^{-12} m²/s for $w/c = 0.5$. According to the results reported by Frederiksen et al. (1997), the D_{RCM28d} values for Swedish SRPC concrete with $w/b = 0.4$ and 0.5 were 14.4×10^{-12} m²/s and 18.6×10^{-12} m²/s, respectively. The D_{RCM6m} values were used in the simple ERFC model and also in the ClinConc model, while both values of D_{RCM6m} and D_{RCM28d} corresponding $t_0 = 0.5$ years (6 months) and 0.077 years (28 days), respectively, were used in the DuraCrete model for comparison. Other parameters were the same as in Section 7.4.1. The results are shown in the following figures.

Case 1: Bridge O 978

Figure 7.8 shows Bridge O 978 over Highway 40 between Borås and Gothenburg, which is subject to heavy and high-speed (>100 km/h) traffic. Figure 7.9 shows the sampling positions. The chloride profiles used for validation were taken from the lower part of the first column against Gothenburg, approximately 3 m from the traffic lane. Labels 'FB'/'FG' and 'MB'/'MG' indicate the sampling positions where the vehicle direction is from ('F') Borås/Gothenburg and where the vehicles pass the column towards ('M') Borås/Gothenburg. The modelled results are shown in Figure 7.10. The simple ERFC model significantly overestimates the chloride ingress, although the

Figure 7.8 A view of Bridge O 978, Landvettermotet (25 years old). Picture taken from the west (courtesy of Lindvall, 2001).

172

D value measured in the 25-year-old concrete was used in the model. The ClinConc model with $w/b = 0.4$ and the DuraCrete model with D_{RCM28d} $w/b = 0.5$ and with D_{RCM6m} $w/b = 0.4$ made relatively good predictions, while the ClinConc model with $w/b = 0.5$ and the DuraCrete model with D_{RCM6m} $w/b = 0.5$ give overestimated predictions. The DuraCrete model with D_{RCM28d} $w/b = 0.4$ underestimated chloride ingress.

Case 2: Bridge O 951

Figure 7.11 shows Bridge O 951 over the highway between Gothenburg and Malmö, which is subject to heavy and high-speed (>100 km/h) traffic. Figure 7.12 shows the sampling positions. The chloride profiles used for validation were taken from the

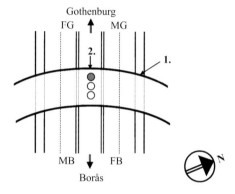

Figure 7.9 Sampling positions on Bridge O 978. The examined column is marked with grey and the locations of the places where the cores have been taken from the side-beam are indicated by '1' and '2' (courtesy of Lindvall, 2001).

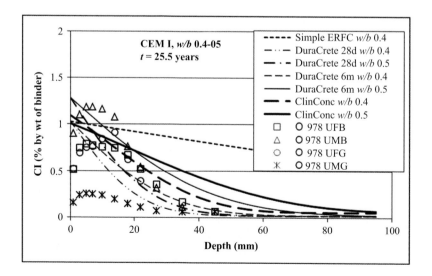

Figure 7.10 Modelled and measured chloride profiles in a 25-year-old road bridge.

Figure 7.11 A view of Bridge O 951, Lindomemotet (27 years old). Picture taken from the south (courtesy of Lindvall, 2001).

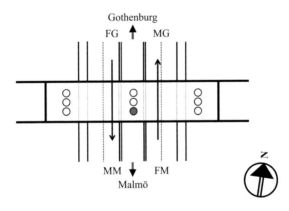

Figure 7.12 Sampling positions on Bridge O 951. The examined column is marked with grey *(based on Lindvall, 2001).*

lower part of the first column against Malmö (marked by 'U'), approximately 2 m from the traffic lane. Marks 'FG'/'FM' and 'MG'/'MM' indicate the sampling positions where the vehicles come from ('F') Gothenburg/Malmö, and where the vehicles pass the column towards ('M') Gothenburg/Malmö, as illustrated in Figure 7.12. The modelled results are shown in Figure 7.13. It can be seen that the ClinConc model predicted the three profiles fairly well but not the highest one, which was taken from the first column on the side receiving splashed water from the vehicles from Malmö. The DuraCrete model with D_{RCM6m} $w/b = 0.5$ also gives a fairly good prediction, but with

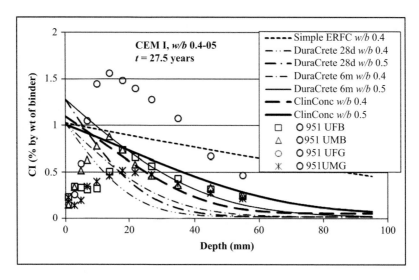

Figure 7.13 Modelled and measured chloride profiles in a 27-year-old road bridge.

Figure 7.14 A view of Bridge O 670, over River Nordre Älv (30 years old). Picture taken from the northwest (courtesy of Lindvall, 2001).

other input parameters it underestimates the chloride ingress in this bridge column. The simple ERFC model still significantly overestimates the chloride ingress.

Case 3: Bridge O 670

Figure 7.14 shows a view of highway bridge O 670, which is subject to heavy and high-speed (>100 km/h) traffic. Figure 7.15 shows the general plan and sampling

175

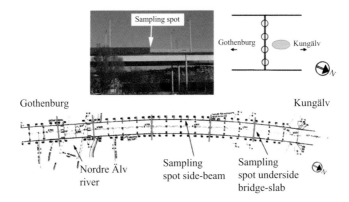

Figure 7.15 General plan showing the sampling spots on Bridge O 670 (based on Lindvall, 2001).

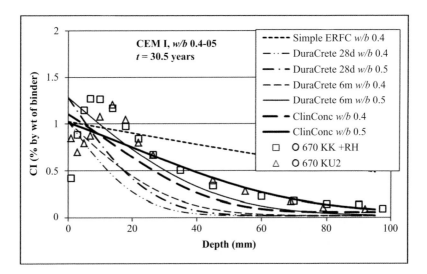

Figure 7.16 Modelled and measured chloride profiles in a 30-year-old road bridge.

spots. Two chloride profiles were used for validation and the results are shown in Figure 7.16. The chloride profile marked 'O 670 KK +RH' was taken from the side beam approximately 3 m from the traffic lane, and the profile marked 'O 670 KU2' was taken from the underside of the pavement slab near the joint where the leakage of water was visible. It can be seen from the results that the ClinConc model gives fairly good predictions for chloride ingress in this concrete bridge. Similar to Figure 7.9, the DuraCrete model with D_{RCM6m} $w/b = 0.5$ also gives fairly good predictions, but with other input parameters it underestimates the chloride ingress. The simple ERFC model again significantly overestimates chloride ingress.

In summary, it can be seen that the simple ERFC model always significantly overestimates chloride ingress. The DuraCrete model with D_{RCM6m} $w/b = 0.5$ gives a

176

good prediction, but with D_{RCM28d} seems to underestimate the chloride ingress in most cases. The reason is probably due to its age factor ($n = 0.65$), which can significantly decrease the apparent diffusion coefficient when t_0 is 28 days (0.077 years) compared with $t_0 = 0.5$ years, as can be seen in equation 7.5. Therefore, the key for the DuraCrete model to achieve a successful prediction is correctly selecting the input parameters. The ClinConc model was previously calibrated against 10-year data from a marine environment (Tang, 2003) and also against site data from the road exposure site (Tang and Utgenannt, 2007). It has been shown from validation against site data from real concrete road bridges after 25–30 years of exposure that this model in general gives fairly good predictions, with predicted profiles close to the measured data.

7.5 Conclusions and recommendations

7.5.1 Conclusions

- The sensitivity analysis shows that, among different parameters, the age factor n is the most sensitive parameter, especially when its value is larger than 0.2. The sensitivity of all the other input parameters is dependent on the ratio C/C_s, except for parameter C_s, which is constantly equal to 1 and independent of other parameters.
- The simple ERFC model significantly overestimates chloride ingress.
- The DuraCrete model, if the input parameters are properly selected (e.g. the value of D_{RCM} measured at $t_0 = 0.5$ years is used), may give reasonably good predictions; otherwise it often underestimates chloride ingress.
- The ClinConc model was previously calibrated against 10-year site data and therefore generally gives fairly good predictions for chloride ingress in old bridges under the de-icing salt environment with heavy traffic at high speed. This is a demonstration of the importance for a model developer to calibrate their prediction model against site data before the model can be applied to the service-life design of concrete structures.

7.5.2 Recommendations

- The DuraCrete model, as described by DuraCrete (Engelund *et al.*, 2000) or Betongföreningen (2007), can be used for the durability design of concrete structures in exposure class XD3 with the input parameters as listed in Table 7.2 for Swedish cement for civil engineering structures (CEM I). With the addition of silica fume, it is suggested that the same input parameters as used

Table 7.2 Input parameters[a] used in the DuraCrete model for XD3.

Parameter	Cement type	Characteristic value	Unit
Environmental factor, $k_{e,cl}$	CEM I	0.68	–
Parameter for surface chloride content, $A_{Cs,cl}$	CEM I	2.57	Cl mass – % of binder
	CEM I + silica fume	3.23	
Age factor, n_{cl}	CEM I	0.65	–

[a]See Engelund *et al.* (2000) and Betongföreningen (2007) for the other input parameters.

for CEM I should be used except for the parameter for surface chloride content, which is higher than that for CEM I. So far, no verified input parameter is available for other types of binder.

- It should be noted that the DuraCrete model with the input parameters as in Table 7.2 is based on verification with site data from road bridges with 25–30 years use in the Gothenburg region. Due to the empirical characteristics of the DuraCrete model, it is uncertain if the results can be extrapolated to service life beyond 30 years.

- The ClinConc model is mechanistically based on chloride transport under submerged conditions. When it is used for non-submerged conditions such as exposure class XD3, some empirical characteristics have to be involved. According to the verification results, this model can predict chloride ingress for both short durations (1–2 years) and long durations (25–30 years). Therefore, it should be safer than the DuraCrete model for extrapolating the results to a certain range of service life (e.g. 50–60 years).

- The guidelines for using the ClinConc model for durability design are given in Appendix 7A.

References

Atlassi, E. (1995) Internal communication, Department of Building Materials, Chalmers University of Technology, Gothenburg.

Andrade, C. *et al.* (2014) Use of resistivity and chloride resistance measurements to assess concrete durability of new panama canal In *Performance-based Specification and Control of Concrete Durability*, Bjegović, D., H. Beushausen and M. Serdar eds.RILEM PRO Zagreb 11–13 June,089, pp. 411–418.

Bamforth, P.B. (2004) *Enhancing reinforced concrete durability: guidance on selecting measures for minimising the risk of corrosion of reinforcement in concrete*, Technical Report No. 61. The Concrete Society, Camberley.

Betongföreningen (2007) *Guidelines for durability design of concrete structures* (in Swedish), Report No. 12. Swedish Concrete Association, Stockholm.

Collepardi, M., A. Marcialis and R. Turriziani (1972) Penetration of chloride ions into cement pastes and concrete. *Journal of American Ceramic Society* **55**(10), 534–535.

Conciatori, D., F. Laferriere and E. Brühwiler (2010) Comprehensive modelling of chloride ion and water ingress into concrete considering thermal and carbonation state for real climate. *Cement and Concrete Research* **40**, 109–118.

Czernin, W. (1964) From *Betonghandbok* (in Swedish), 2nd edn, C. Ljungkrantz *et al.* (eds). AB Svensk Byggtjänst and Cementa AB, Stockholm, Sweden, 1994, 282.

Engelund, S., C. Edvardsen and L. Mohr (2000) *General guidelines for durability design and redesign, DuraCrete: probabilistic performance based durability design of concrete structures*, EU-Project (Brite EuRam III) No. BE95-1347, Report R 15.

fib (2006) Model code for service-life design, fib bulletin 34, Federation International du Beton, Lausanne, Switzerland, 1st edn.

Frederiksen, J.M. and M. Geiker (2000) On an empirical model for estimation of chloride ingress into concrete. *Proceedings of 2nd International RILEM Workshop on Testing and Modelling the Chloride Ingress into Concrete*, September 2000, Paris, RILEM PRO 19, pp. 355–371.

Frederiksen, J.M. and M. Geiker (2008) Chloride ingress prediction – Part 1 & Part 2. *Proceeding of International RILEM Symposium on Concrete Modelling – ConMod '08*, RILEM PRO 58, pp. 275–290.

Frederiksen, J.M., L.-O. Nilsson, E. Poulsen, P. Sandberg, L. Tang and A. Andersen (1996) *HETEK – a system for estimation of chloride ingress into concrete. Theoretical background*, Report No. 83. Danish Road Directorate, Copenhagen.

Frederiksen, J.M., H.E. Sørensen, A. Andersen and O. Klinghoffer (1997) *HETEK – the effect of the w/c ratio on chloride transport into concrete – Immersion, migration and resistivity tests*. Report No. 54. Danish Road Directorate, Copenhagen.

Gang, X., L. Yun-pan, S. Yi-Biao and X. Ke (2015) Chloride ion transport mechanism in concrete due to wetting and drying cycles. *Structural Concrete* **2**, 289–296.

Hosokawa, Y., K. Yamada, B. Johannesson and L.-O. Nilsson (2008) A development of a multi-species mass transport model considering thermodynamic phase equilibrium. *Proceeding of International RILEM Symposium on Concrete Modelling – ConMod '08*, RILEM PRO 58, pp. 543–550.

Jin, W.-L. and C. Xu (2010) *Service life prediction of concrete marine structures*. In 2nd International Symposium on Service Life Design for Infrastructures, K. van Breugel, K., G. Ye and Y. Yuan eds. RILEM Publications, Paris, 339–350.

Johannesson, B. (2000) *Transport and sorption phenomena in concrete and other porous media*. PhD thesis, Report TVBM-1019. Division of Building Materials, Lund Institute of Technology, Lund.

Lindvall, A. (2001) *Environmental actions and response – reinforced concrete structures exposed in road and marine environments*. Licentiate thesis, Department of Building Materials, Chalmers University of Technology, Publication No. P-01:3, Gothenburg, Sweden.

Maekawa, K., T. Ishida and T. Kishi (2003) Multi-scale modeling of concrete performance. Integrated material and structural mechanics. *Journal of Advanced Concrete Technology* **1**(2), 91–126.

Meijers, S.J.H. (2003) *Computational modelling of chloride ingress in concrete*. PhD thesis, Delft University, Delft.

Meijers, S.J.H., J.M.J.M. Bijen, R. de Borst and A.A. Fraaij (2005) Computational results of a model for chloride ingress in concrete including capillary suction, drying-wetting cycles and carbonation. *Materials and Structures* **38**, 145–154.

Mejlbro, L. (1996) The complete solution of Fick's second law of diffusion with time-dependent diffusion coefficient and surface concentration. In *Durability of Concrete in Saline Environment*, CEMENTA AB, Danderyd, Sweden, pp. 127–158.

Nguyen, T.Q., V. Baroghel-Bouny, P. Dangla and P. Belin (2006) Multi-level modelling of chloride ingress into saturated concrete. *Proceedings of International RILEM Workshop on Performance Based Evaluation and Indicators for Concrete Durability*, 19–21 March, Madrid, Spain.

Nordtest (1999) *NT BUILD 492 – Concrete, mortar and cement based repair materials: chloride migration coefficient from non-steady state migration experiments*. NORDTEST, Espoo.

Petre-Lazar, I., L. Abdou, C. Franco and I. Sadri (2003) THI: a physical model for estimating the coupled transport of heat, moisture and chloride ions in concrete. *Proceedings of 2nd International RILEM Workshop on Life Prediction and Aging Management of Concrete Structures*, 5–6 May, Paris, pp. 55–69.

Poulsen, E. (1993) On a model of chloride ingress into concrete having time-dependent diffusion coefficient. In *Chloride Penetration into Concrete Structures - Nordic Miniseminar*. Division of Building Materials, Chalmers University of Technology, Publication P-93:1, Gothenburg, pp. 298–309.

Samson, E., J. Marchand, L. Robert and J.P. Bournazel (1999) Modeling the mechanisms of ion diffusion in porous media. *International Journal for Numerical Methods in Engineering* **46**, 2043–2060.

Sergi, W., S.W. Yu and C.L. Page (1992) Diffusion of chloride and hydroxyl ions in cementitious materials exposed to a saline environment. *Magazine of Concrete Research* **44**(158), 63–69.

Shi, X. *et al.* (2012) Durability of steel reinforced concrete in chloride environments: an overview. *Construction and Building Materials* **30**, 125–138.

Tang, L. (1996) *Chloride transport in concrete – measurement and prediction.* Doctoral thesis, Department of Building Materials, Chalmers Universities of Technology, Publication P-96:6, Gothenburg.

Tang, L. (2003) *Chloride ingress in concrete exposed to marine environment – field data up to 10 years exposure*, SP Report 2003:16, SP Swedish National Testing and Research Institute, Borås.

Tang, L. (2005) *WP5 Report. Final evaluation of test methods, ChlorTest: Resistance of concrete to chloride ingress – from laboratory tests to in-field performance.* EU-Project (5th FP GROWTH) G6RD-CT-2002-00855, Deliverables 16–19.

Tang, L. (2006) Service-life prediction based on the rapid migration test and the ClinConc model. In *RILEM Proceedings PRO 047: Performance Based Evaluation and Indicators for Concrete Durability*, Baroghel-Bouny, V. *et al.* (eds), RILEM Publications, pp. 157–164.

Tang, L. (2008) Engineering expression of the ClinConc model for prediction of free and total chloride ingress in submerged marine concrete. *Cement and Concrete Research* **38**(8–9), 1092–1097.

Tang, L. and L.-O. Nilsson (1993) Chloride binding capacity and binding isotherms of OPC pastes and mortars. *Cement and Concrete Research* **23**, 347–353.

Tang, L. and L.-O. Nilsson (1994) A numerical method for prediction of chloride penetration into concrete structures. In *The Modelling of Microstructure and its Potential for Studying Transport Properties and Durability.* Kluwer Academic, pp. 539–552.

Tang, L. and L.-O. Nilsson (1996) Service life prediction for concrete structures under seawater by numerical approach. *Proceedings of the 7th International Conference on the Durability of Building Materials and Components*, 19–23 May 1996. Stockholm, E & FN Spon, pp. 97–106.

Tang, L. and P. Utgenannt (2007) *Chloride ingress and reinforcement corrosion in concrete under de-icing highway environment - a study after 10 years' field exposure*, SP Report 2007:76. SP Technical Research Institute of Sweden, Borås.

Tang, L., L.-O. Nilsson and P.A.M. Basheer (2011) *Resistance of concrete to chloride ingress – testing and modelling.* Taylor & Francis, London.

Tang, L., P. Utgenannt, A. Lindvall and D. Boubitsas (2012) *Validation of models and test methods for assessment of durability of concrete structures in the road environment*, CBI Report 2:2012. Swedish Cement and Concrete Research Institute, Stockholm.

Thomas, M.D.A. and Bentz (2001) *Life 365 – Computer program for predicting the service life and life-cycle costs of reinforced concrete structures exposed to chlorides*, User Manual (version 1.0.0), presented at the Nordic Mini Seminar & fib TG 5.5 Meeting, 22–23 May, Gothenburg, Sweden.

Truc, O., J.P. Ollivier and L.-O. Nilsson (2000) Numerical simulation of multi-species diffusion. *Materials and Structures* **33**(122), 566–573.

Utgenannt, P. (2004) *The influence of ageing on the salt-frost resistance of concrete*, PhD thesis, TVBM-1021. Division of Building Materials, Lund Institute of Technology, Lund, Sweden.

Xu, A. (1992) *Structure of hardened cement-fly ash systems and their related properties*, Doctoral thesis, Publication 92:7. Department of Building Materials, Chalmers University of Technology, Gothenburg, Sweden.

Appendix 7A
Guidelines for application of the ClinConc model in durability design

The guidelines are based on Tang's engineering approach (Tang, 2006, 2008), under the assumption of semi-infinite boundary condition.

7A.1 Modelling of free chloride ingress

The first step in predictions with the ClinConc model is to determine the free chloride content in the concrete at depth x. This is carried out with the following expression:

$$\frac{c-c_i}{\gamma_{Cs}c_s-c_i} = 1-\mathrm{erf}\left(\frac{x}{2\sqrt{\dfrac{\gamma_D \xi_D D_{6m}}{1-n}\cdot\left(t'_{6m}\right)^n\cdot\left[\left(1+\dfrac{t'_{ex}}{t}\right)^{1-n}-\left(\dfrac{t'_{ex}}{t}\right)^{1-n}\right]\cdot t}}\right) \quad (7A.1)$$

where c, c_s and c_i are the concentration of free chlorides in the pore solution at depth x, at the surface of the concrete and initially in the concrete, respectively, γ_{Cs} is the partial factor taking into account the variation of chloride concentration in the exposure environment, D_{6m} is the diffusion coefficient measured by the RCM test at the age of t'_{6m}, γ_D is the partial factor taking into account the variation of the measured chloride diffusion coefficient, ξ_D is the factor bridging the laboratory measured D_{6m} to the initial apparent diffusion coefficient for the actual exposure environment, n is the age factor accounting for the decrease of diffusivity with age, t'_{ex} is the age of the concrete at the start of exposure, and t is the duration of exposure.

Factor ξ_D is given by the following expression:

$$\xi_D = \frac{\left(0.8\cdot a_t^2 - 2\cdot a_t + 2.5\right)\cdot\left(1+0.59\cdot K_{b6m}\right)\cdot e^{\frac{E_D}{R}\left(\frac{1}{293}-\frac{1}{T}\right)}}{1+k_{OH6m}\cdot K_{b6m}\cdot f_b\cdot\beta_b\cdot\left(\dfrac{c_s}{35.45}\right)^{\beta_b-1}\cdot e^{\frac{E_b}{R}\left(\frac{1}{T}-\frac{1}{293}\right)}}\cdot k_D \quad (7A.2)$$

where a_t is a factor that describes how the chloride binding changes over time, f_b and β_b are chloride binding constants, E_D and E_b are activation energy chloride diffusion and binding, respectively, k_D is the expansion factor depending on the type of binder and water–binder ratio, and k_{OH6m} and K_{b6m} are factors accounting for the effects of hydroxide concentration in the pore solution, gel content and water accessible porosity at age t'_{6m}. Based on the limited literature data, the activation energy $E_D = 42{,}000$ J/mol and $E_b = 40{,}000$ J/mol are adopted (Tang and Nilsson, 1996). Furthermore,

$$k_{OH6m} = e^{0.59\left(1-\frac{0.043}{[OH]_{6m}}\right)} \quad (7A.3)$$

181

and

$$K_{b6m} = \frac{W_{gel6m}}{1000\varepsilon_{6m}} \qquad (7A.4)$$

where $[OH]_{6m}$, W_{gel6m} and ε_{6m} are the hydroxide concentration in mol/m³$_{pore\text{-}solution}$, gel content in kg/m³$_{concrete}$ and water accessible porosity at age t'_{6m}.

According to experience obtained from 10 years of exposure in seawater at the Swedish west coast (Tang, 2003), the expansion factor k_D can be estimated by

$$k_D = \begin{cases} 1+8(0.4-w/b)+7SF+3800(SF \cdot FA)\cdot(SF+FA) & 0.25 \leq w/b \leq 0.4 \\ 1 & w/b > 0.4 \end{cases} \qquad (7A.5)$$

where w/b is the water–binder ratio, SF and FA are the mass fraction of silica fume and fly ash to the total binder, respectively. This expansion factor describes the ratio of the diffusion coefficient in the site concrete to that in the laboratory test.

Similarly, the binding factor a_t can be estimated by

$$a_t = \begin{cases} 0.36+1.4(0.4-w/b)+0.4SF+38(SF \cdot FA)\cdot(SF+FA) & 0.25 \leq w/b \leq 0.4 \\ 0.36+1.4(0.4-w/b) & 0.4 < w/b \leq 0.6 \end{cases} \qquad (7A.6)$$

In equations 7A.5 and 7A.6, the effects of pozzolanic additions become insignificant when $w/c > 0.4$, probably due to there being a sufficient volume of the capillary network in the concrete with sufficient high w/c (>0.4). This capillary volume can release stresses from the post-hydration of pozzolanic additions and the capillary network can give sufficient paths for the gel to bind the penetrated free chlorides. Owing to the complicated preparation of test samples for chloride-binding measurements, the values of chloride-binding constants, f_b and β_b, are limited. Based on the data reported by Tang (1996), and the practical applications in the modelling work (Tang, 2003), the values $f_b = 3.6$ and $\beta_b = 0.38$ are applicable to concrete with Portland cement or silica fume. For binders with fly ash (FA) and slag (BFS) the following equations may be used to estimate these constants:

For fly ash: $\qquad\qquad f_b = 3.6 + 7FA$ and $\beta_b = 0.38 - 0.3FA \qquad\qquad$ (7A.7)

For slag: $\qquad\qquad\;\; f_b = 3.6 + 3.5BFS$ and $\beta_b = 0.38 - 0.14BFS \qquad$ (7A.8)

where BSF is the mass fraction of slag to the total binder.

The age factor n is, according to the numerical simulation made by Tang (2008), mainly attributed to the increase in chloride binding through parameter a_t, and can be expressed by the following regression equation:

$$n = -0.45a_t^2 + 0.66a_t + 0.02 \qquad (7A.9)$$

It should be noted that the above equation was based on the submerged environment, where the chloride solution is constantly in contact with the concrete surface. When the model is used for the atmospheric zone or the road environment, certain modifications are needed. In this study, the following modifications were made: 1) surface free chloride concentration c_s was estimated based on the short-term exposure experiment; 2) the time-dependent factor of chloride binding, a_t, was considered to be one-third that of the submerged environment; and 3) the age factor due to drying, $n_{dry} = 0.5$, was added to equation 7A.9:

$$n = \left(-0.45a_t^2 + 0.66a_t + 0.02\right) + n_{dry} \qquad (7A.9')$$

Obviously, these modifications are empirical. Further numerical simulations are needed in order to 'mechanistically' clarify these modifications.

7A.2 Calculation of total chloride content

The total chloride content, c_{tot}, or expressed with a capital letter C to differentiate it from the free chloride concentration c, is basically the sum of the bound chloride, c_b, and free chloride, c. However, this requires that the relationship between the free and total chloride content (i.e. a chloride binding isotherm) is known. In cases when this relationship is not known the following expression may be used:

$$C = \frac{\varepsilon \cdot (c_b + c)}{B_c} \times 100 \quad \text{(as mass\% of binder)} \qquad (7A.10)$$

where ε is the water-accessible porosity at the age after exposure, B_c is the cementitious binder content (in kg/m$^3_{concrete}$) and

$$c_b = f_t \cdot k_{OH6m} \cdot K_{b6m} \cdot f_b \cdot c^{\beta_b} \cdot e^{\frac{E_b}{R}\left(\frac{1}{T} - \frac{1}{293}\right)} \text{(in g/l)} \qquad (7A.11)$$

where f_t is a factor accounting for the time dependency of chloride binding and can be calculated by the following equation:

$$f_t = a_t \ln\left(\frac{c - c_i}{c_s - c_i} \cdot t + 0.5\right) + 1 \qquad (7A.12)$$

The other parameters, such as porosity, gel content and hydroxide concentration, are dependent on individual mixture proportions and can be estimated using well-established equations in the concrete handbooks or the equations in Section 7A.5.

7A.3 Prediction of service life

If the threshold chloride concentration c_{cr} or content C_{cr} is known, the ClinConc model can be used for the prediction of service life. In the former case, the free chloride concentration c in equation 7A.1 is simply replaced with c_{cr}. If the service life t_L is specified, the cover thickness x_c can be calculated from the following equation:

$$x_c = 2\sqrt{\frac{\gamma_D \xi_D D_{6m}}{1-n} \cdot \left(\frac{t'_{6m}}{t_L}\right)^n \cdot \left[\left(1+\frac{t'_{ex}}{t_L}\right)^{1-n} - \left(\frac{t'_{ex}}{t_L}\right)^{1-n}\right] \cdot t_L \cdot \text{erf}^{-1}\left(1 - \frac{c - c_i}{\gamma_{Cs} c_s - c_i}\right)} \qquad (7A.13)$$

If the cover thickness x_c is specified, this equation can be rearranged as

$$\left(\frac{t'_{6m}}{t_L}\right)^n \cdot \left[\left(1+\frac{t'_{ex}}{t_L}\right)^{1-n} - \left(\frac{t'_{ex}}{t_L}\right)^{1-n}\right] \cdot t_L = \frac{1-n}{4\gamma_D \xi_D D_{6m}} \cdot \left[\frac{x_c}{\mathrm{erf}^{-1}\left(1-\dfrac{c-c_i}{\gamma_{Cs}c_s - c_i}\right)}\right]^2 \qquad (7A.14)$$

In this case of t_L, equation 7A.14 can be solved with the use of 'Solver' in MS Excel.

When the threshold chloride is expressed by C_{cr} in mass% of binder, the following equation can be used, again with the help of 'Solver' in MS Excel, to convert C_{cr} (mass% of binder) to c_{cr} (g/l) in order to utilise equations 7A.13 and 7A.14:

$$k_{OH6m} \cdot K_{b6m} \cdot f_b \cdot c_{cr}^{B_b} \cdot e^{\frac{E_b}{R}\left(\frac{1}{T} - \frac{1}{293}\right)} + c_{cr} = \frac{B_c C_{cr}}{100\varepsilon} \qquad (7A.15)$$

7A.4 Consideration of uncertainty

Similar to the DuraCrete model, a partial factor can be added to cover the uncertainty in some of the input parameters. Based on current knowledge, two partial factors, that is, γ_{Cs} and γ_D, as in equations 7A.1, 7A.13 or 7A.14, are considered. Their values can be determined by multiplying the standard deviation or coefficient of variance (*COV*) of the respective measurement with a coverage factor of 2, implying a 95% confidence, that is,

$$\gamma = \frac{m+2s}{m} = 1 + 2 \times \frac{s}{m} = 1 + 2 \times COV \qquad (7A.16)$$

For the Swedish road environment this comprises a lack of information about the surface free chloride concentration and its variation. Based on the verification presented in Sections 7.5.1.1 and 7.5.1.2, it is rational to adopt the value $\gamma_{Cs} = 1.5$. For the diffusion coefficient measured by the RCM test, the reproducibility *COV* in a European interlaboratory evaluation is 23% (Tang, 2005). The value of $\gamma_D = 1.46$ is therefore suggested.

7A.5 Equations for the parameters related to the concrete

If no standard equation or handbook is available the following equations can be used for estimation of the parameters related to the concrete.

Degree of hydration

$$\alpha_h = \alpha_{max} \cdot \exp\left[-A_\alpha \left[\ln(t)\right]^{-B_\alpha}\right] \qquad (7A.16)$$

where α_{max} is the maximum degree of hydration,

$$\alpha_{max} = \begin{cases} 1 & {}^w\!/\!_c \geq 0.39 \\ \dfrac{{}^w\!/\!_c}{0.39} & {}^w\!/\!_c < 0.39 \end{cases} \qquad (7A.17)$$

A_α and B_α are constants,

$$B_\alpha = \frac{\ln\left[\dfrac{\ln\left(\dfrac{\alpha_{h2}}{\alpha_{max}}\right)}{\ln\left(\dfrac{\alpha_{h1}}{\alpha_{max}}\right)}\right]}{\ln\left[\dfrac{\ln(t_1)}{\ln(t_2)}\right]} \text{ and } A_\alpha = -\frac{\ln\left(\dfrac{\alpha_{h1}}{\alpha_{max}}\right)}{\left[\ln(t_1)\right]^{-B_\alpha}}$$

For OPC, at an age of 2.5 years,

$$\frac{\alpha_{h2}}{\alpha_{max}} = \begin{cases} 1 & w/c \geq 0.625 \\ 1.265 \times \sqrt{w/c} & w/c < 0.625 \end{cases} \quad \text{(based on the data from Czernin, 1964)}$$

At an early age (e.g. 1 day),

$$\alpha_{h1} = 0.48 \times \sqrt{w/c} \cdot \exp\left[-\frac{E_\alpha}{R}\left(\frac{1}{T} - \frac{1}{293}\right)\right] \quad \text{(based on the data from Atlassi, 1995)}$$

where $E_\alpha = 36{,}000$ J/mol when $T \geq 293$ K, or $E_\alpha = 65{,}000$ J/mol when $T < 293$ K.

Non-evaporable water

$W_n = 0.25\alpha_h \cdot C - 0.34k_{SF} \cdot SF + 0.25k_{FA} \cdot FA + 0.25k_{SL} \cdot SL$ (in kg/m$^3_{concrete}$)

where C is the cement content (in kg/m$^3_{concrete}$), k_{SF} is the silica fume coefficient ($k_{SF} = 0.0212t^{0.44}$; max 0.85, based on data from Atlassi, 1995), SF is the silica fume content (in kg/m$^3_{concrete}$), k_{FL} is the fly ash coefficient ($k_{FL} = 0.0286t^{0.286}$; max 0.5, based on data from Xu, 1992), FA is the fly ash content (in kg/m$^3_{concrete}$), k_{SL} is the slag coefficient and is assumed equal to a_h, and SL is the slag content (in kg/m$^3_{concrete}$).

The hydration age t in the above equations is in hours.

Gel quantity

$$W_{gel} = \alpha_h \cdot C + k_{SF} \cdot SF + k_{FA} \cdot FA + k_{SL} \cdot SL + W_n \text{ (in kg/m}^3_{concrete}) \quad (7A.18)$$

Diffusible porosity under the capillary saturated condition

$$\varepsilon = 1 - \sum \frac{m}{\rho} - 0.75W_n - \varepsilon_{air} \quad (7A.19)$$

where m and ρ are the mass and specific density of solid materials in concrete, respectively, and ε_{air} is the porosity of air voids.

Hydroxide concentration in the pore solution

$$[OH] = \frac{B_c}{\varepsilon} \cdot \frac{2 \times (Na_2O)_{eq}}{62} \text{ (in kmol/m}^3_{solution} \text{ or mol/l)} \quad (7A.20)$$

where the content of equivalent Na_2O is

$$\left(Na_2O\right)_{eq} = 0.66 \times \left(K_2O\right) + \left(Na_2O\right) \ \text{(in kg/kg}_{binder}) \qquad (7A.21)$$

7A.6 Suggested parameters for the Swedish road environment

There is a lack of a valid measurement method for surface free chloride concentration in the de-icing salt road environment. Based on verification, a value of $c_s = 1.5$ g/l can be suggested. Based on the measurement data at the site by Highway 40 outside Borås in southwest Sweden, the annual average temperature $T = 10°C$ is suggested at least for the western region. For southern regions, a slightly higher value may be used. For the other regions, the use of $T = 10°C$ would be on the conservative side.

Chapter 8
Evolution of corrosion parameters in a buried pilot nuclear waste container in El Cabril

C. Andrade, Samuel Briz, P. Zuloaga, M. Ordoñez and F. Jimenez

The need for long-term storage of radioactive waste involves several challenges due to the lack of records regarding the prolonged durabilities of modern materials. The El Cabril repository in Cordoba, Spain, for low and medium radioactivity waste has a design life of 300–500 years, which exceeds present experience about the use of structural concrete, which dates from about the beginning of the 20th century. This lack of records on structures older than 100–150 years calls for more specific studies to anticipate long-term behaviour. In order to study the real on-site aging of concrete, Enresa has undertaken a monitoring programme in collaboration with IETcc and Geocisa through the installation of permanent sensors in a pilot container buried under realistic conditions. The container was provided with a set of sensors (installed in 1995) to monitor durability parameters. This chapter presents the evolution from 1995 of these corrosion parameters. The sensors, by means of non-destructive tests (NDTs), measure temperature, corrosion rate, corrosion potential, electrical resistivity, local concrete strain and oxygen availability. Relations between the seasonal changes, depth of burying and the corrosion parameters are also presented. The results provided by the sensors indicate that the temperature is a very relevant variable influencing the measurements and that as the burial depth increases, the temperature becomes more stable, changing less with the exterior changes. Also the extreme values of temperature decrease with buried depth. Apparent activation energies were calculated in the case of local strain, and electrical resistivity. Long-term predictions were made with values of temperature in the range 19–21°C.

8.1 Introduction

The El Cabril repository for low and medium radioactive waste storage is based on a multibarrier insulation concept (Figures 8.1 and 8.2). Concrete is part of the engineering barrier used to minimise water ingress and percolation through the waste and thus to minimise radionuclide leaching and transport to the geosphere, biosphere and finally to humans. Isolation of radionuclides is the main function of the engineered barriers, which is interpreted in terms of dose restriction (e.g. 0.1 mSv/year to any individual in a critical group from the public).

Figure 8.1 shows the main cement-based construction materials needed in an engineering barrier in the repository of El Cabril for low and medium radioactive wastes (Zuloaga, 2010). The construction comprises concrete vaults, concrete containers, primary package cementitious matrices and cement grout backfilling of the concrete

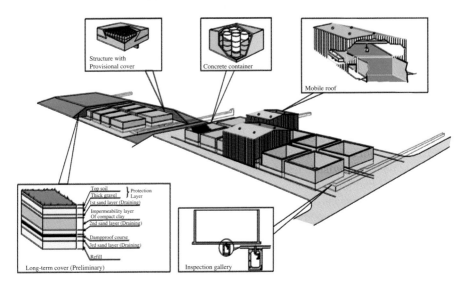

Figure 8.1 General view of El Cabril facility.

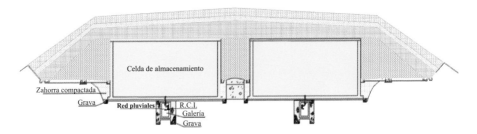

Figure 8.2 Vaults with a final covering of ground.

containers. Figure 8.2 shows a final cross-section of the vaults (using the multibarrier concept) when buried underground. The vaults have an inspection gallery underneath that detects leakage.

The construction has a design life of 300 years, with low maintenance requirements divided into three main periods: operations during construction, observation where a minimum of monitoring is carried out, and post-observation where minimal control is expected. The durability required for such extended periods makes it necessary to have confidence in the construction. Confidence in the facility is supported by modelling, which gives robust solutions, plus excellent knowledge of the degradation processes.

Scenarios were developed to identify a wide range of deterioration processes in cement-based materials caused by environmental site conditions (Andrade *et al.*, 2006). Those with the highest probability were carbonation (during the operational phase only), water permeation (leaching) and reinforcement corrosion. More unlikely causes that could appear are bio-attacks. Chloride attack of the reinforcement is also probable, as chlorides are contained in the drums as part of the waste from medical analysis.

A research programme was launched by Enresa in cooperation with Instituto Eduardo Torroja (IETCC-CSIC), initially focusing on optimising the durability of the engineered barriers, taking into account all the potential degradation mechanisms. In a second part the goal was service life assessment: models, ranging from simple models to sophisticated models, were developed to assess the service life of the barriers and the characterisation of real samples was enhanced to obtain additional data to verify those models. Since the initial operation, the research and observation programme has included specimen follow-ups from fabricated containers and real packages and from the construction of the vaults, stored in different conditions, measurement of the permeability to gas of the covering, and measurement of corrosion parameters (corrosion potential and velocity) in real vault walls (Andrade *et al.*, 2000, 2006). The current programme consists of the following:

- Control tests (mechanical, sorption, porosimetry) on a yearly basis
- Characterisation of materials, paying attention to permeability, thermal properties, unsaturated permeability, distribution coefficient retention curves plus concentration profiles to verify models
- Durability tests, including leaching (surface and flow through), chloride depassivation thresholds, sulfate resistance and sulfate affection thresholds and freeze–thaw
- Modelling, including thermohydromechanical models and connected reactive transport models to consider non-isothermal conditions and capillary water absorption
- Data collection at the instrumented pilot container test bed, as it was decided not to dismount the pilot container and update the data collection system
- Monitoring of the disposal vaults through gas permeability and corrosion parameter measurements every six months on the last vault to be filled with waste, and through the installation of thermohygrometric instrumentation in two vaults, one on the outer surface and one on the interior surface on the pile of waste containers, with the cables passing through the drainage port

This chapter presents the evolution of the corrosion parameters recorded in the buried pilot container, in conditions representative of the actual disposal vaults, as this is an extreme environmental condition for which almost no data has, so far, been collected.

8.2 Pilot container and its instrumentation

The containers were prefabricated (Andrade *et al.*, 2000, 2006) and cast by means of robotic tools and finally steam-cured at temperatures below 60°C. The typical characteristic concrete strengths at 28 days were 50–60 MPa. The pilot container was fabricated in 1994 and provided with instruments during fabrication by placing sensors (27 sets of electrodes) in position before concreting. Several views of the container and its instrumentation are shown in Figure 8.3. Figure 8.4 shows the position of all groups of sensors (electrodes).

The pilot container was filled with drums (Figure 8.4, right) that had been instrumented with two groups of electrodes at two levels by previously removing the black paint and making the metallic connection directly to the base metal of the drum. After

189

Figure 8.3 Left top: Sensors attached to the reinforcement. Left bottom: Cables of the sensors. Middle top: Instrumented drum before being placed inside the container. Middle bottom: Drums in their final position. Right top: Container being moved after fabrication. Right bottom: Buried container with its access hole at the left of the image.

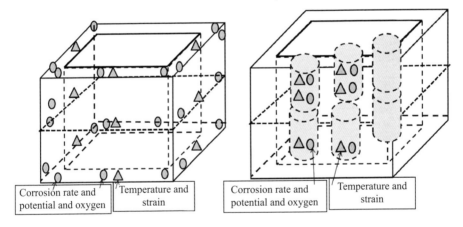

| Corrosion rate and potential and oxygen | Temperature and strain | Corrosion rate and potential and oxygen | Temperature and strain |

Figure 8.4 Positioning of the groups of sensors in the container and drums.

placing the six drums in the container it was finally covered with a cap and the container was placed into a receptacle that simulated the vault. To reproduce the real conditions, the gap between the vault and the container was filled with sand.

The container arrangement is shown in Figure 8.5. Two receptacles were prepared: the container was placed in one and the corrosion data logger in the other (Geologger measurement system), together with a specimen taken as a reference with all the embedded sensors (channel 0) (Figure 8.5, right). There was a main top access to the chamber (Figure 8.5, left) where the data logger (Geologger) was placed beside the container. The cables from the back of the Geologger (Figure 8.5, centre) were

190

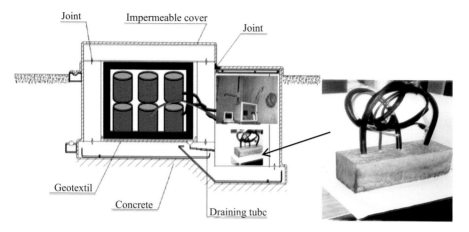

Figure 8.5 Simulated vault where the container is located, and the side receptacle where the Geologger and the reference specimen with the set of embedded sensors (channel 0) are located.

passed to the receptacle, where the container was located, through a hole in the lateral wall of the vault.

The Geologger was a potentiotat-galvanostat with 50 independent measurement channels and was specifically designed for this application. The system was installed in 1994 and a new system for the same purpose was installed in 2007. The parameters measured through the Geologger were temperature, local concrete strain, corrosion potential, resistivity, oxygen availability and corrosion rate. The sensors or electrodes used for the measurements are shown in Figure 8.6. Measurements were made four times per day to capture the most extreme conditions and so as not to record unnecessary data.

8.3 Techniques

To measure the corrosion potential and corrosion rate, direct metallic contact was made with the main rebar of the container or by removing the black paint surface of the drum. The rebar and drum acted as working electrodes. The reference electrode was constructed from Mn/MnO_2 (Figure 8.6). For the corrosion rate I_{corr}, the method used was to measure the polarization resistance R_p (Andrade and Gónzales, 1978; Rilem, 2004) in galvanostatic mode. The mathematical expression for this is

$$R_p = \left(\frac{\Delta E}{\Delta I} \right)_{E < 20 mV} \rightarrow I_{corr} = \frac{B}{R_p} \qquad (8.1)$$

where ΔE is the shift in potential after application of a current pulse ΔI, and B is a constant whose average value is 26 mV.

As the surface of the working electrode is larger than the counter electrode (small disk) a technique based on the measurement of the slope of the transient pulse after application of a step of current was used. This technique is not as accurate as the 'sensorised guard ring' approach (Feliú et al., 1990). However, as the technique is continuously recorded, any scatter or incorrect measurements can be identified easily.

191

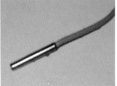

Figure 8.6 Electrochemical sensors embedded in the pilot container. From left to right: Corrosion rate and oxygen, resistivity, corrosion potential and temperature.

However, it has the advantage of being very fast, and it inflicts a very small disturbance on the system. Clearly, a method based on the use of a guard ring around the small counter disk entails difficulties when operated embedded in concrete. Thus, an alternative methodology had to be used, in spite of a loss of accuracy.

The resistivity (Feliú *et al.*, 1996; Polder *et al.*, 2000) was measured using the current interruption method from a galvanostatic pulse. The oxygen flow at the rebar level was measured by applying a cathodic constant potential of about –750 mV (SCE–saturated calomel electrode) (Gjørv and Vennesland, 1983), and measuring the change of current, which relates to the change in the reduction of oxygen. Its availability is measured through Faraday's law:

$$\frac{mole\ O_2}{m^2.\ s} = \frac{I}{4F} \tag{8.2}$$

where I is the current density at –750 mV and F is Faraday's constant.

8.4 Results

Among the 27 groups of sensors installed, fewer than 15% of them failed. Some sensors completely failed from the start, but other sensors failed temporarily but later recovered. The remaining sensors still show a good response 20 years after installation.

8.4.1 Relative humidity in the receptacle

Sensors for relative humidity (RH) and temperature were placed in the receptacle. Their values up to 2009 are given in Figure 8.7. The RH is high although stable over the years, following the seasonal variations of temperature between 10°C and 35°C. The highest values of RH (~90%) were reached in the winter and the lowest (~60–70%) in the summer. The high values of RH are attributed to the room being enclosed by walls of concrete holding the moisture in equilibrium. It can be inferred that the space where the container is placed has the same regime of RH and temperature variations. Thus it can be expected that the concrete gel pores will be saturated, but not the concrete capillary pores.

8.4.2 Sensor 0 (reference specimen) temperature

The reference specimen was placed in the side receptacle (Figure 8.5) together with the Geologger and had the same environment as the side receptacle. The temperature followed the seasonal variations of the air in the side receptacle. Figure 8.8, right, shows a comparison of the temperature in the air and in the specimens. Around 2.5°C is the ordinate at the origin of the linear regression line between both temperatures. It can be seen that they are very similar, although the temperature of the side receptacle is a little higher.

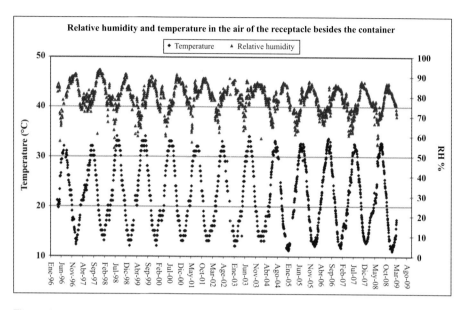

Figure 8.7 Variation of temperature (lower data, in blue) and relative humidity (upper data, in red) in the receptacle beside the container.

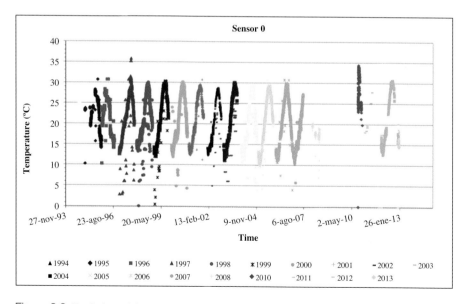

Figure 8.8 Evolution of the temperature inside the specimen of the reference sensor (0), located in the side receptacle.

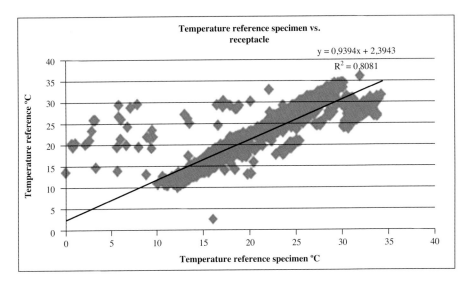

Figure 8.8 (*Continued*)

8.4.3 Sensors embedded in the container walls and attached to the drums

Monitoring of the parameters of the sensors can be visualised as in the example in Figure 8.9. From this figure it is possible to deduce that the resistivity evolves due to seasonal changes following temperature development.

For the remainder of the sensors their changes with time up to 2013 are shown in Figures 8.10 to 8.15. From these figures it is possible to deduce that the temperature changes due to seasonal changes are the most influential factors for the recorded trends. The changes in temperature in all sensors and in the reference specimen are depicted in Figure 8.10. These records indicate that, although the average values are similar at all levels, the maximum temperature is higher in the top part of the container, reaching around 37°C, while at the lower level the maximum is around 32°C. This behaviour is less pronounced in the drums inside the container where the temperatures are roughly similar in the top and bottom of them.

Figure 8.11 shows the evolution of local strains, which initially indicate fast shrinkage, and then a slower increase with time. The values appear to stabilise after a number of years. One value shows an expansion that was produced due to the accidental entrance of moisture into the container.

Regarding oxygen evolution (Figure 8.12), the values are initially relatively high but then suffer an important reduction during the first few years until, later, the cathodic current becomes almost zero or even positive up to 2007. This behaviour would indicate that an anaerobic ambient is being produced, but the values of corrosion potential (Figure 8.13) do not show this trend. However, the corrosion potential evolves slowly but steadily towards more positive values, indicating that oxygen is not

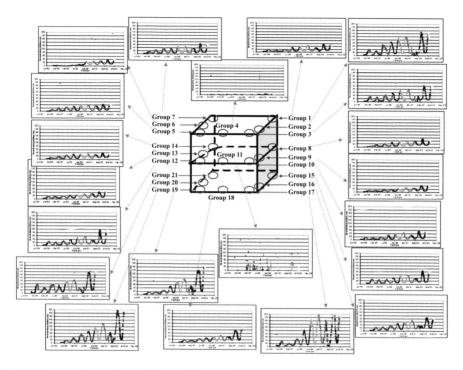

Figure 8.9 Evolution of the resistivity of all the sensors in the container walls.

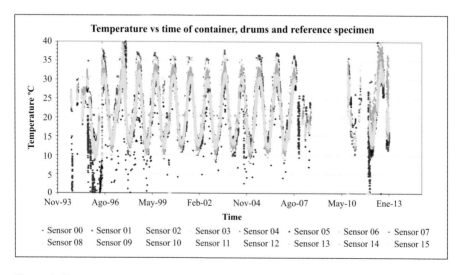

Figure 8.10 Evolution over time of the temperature in the concrete, the mortar and the reference specimen.

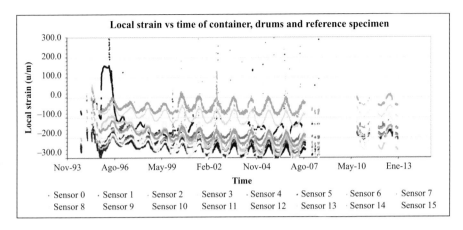

Figure 8.11 Evolution over time of the local strains in the concrete, the mortar and the reference specimen.

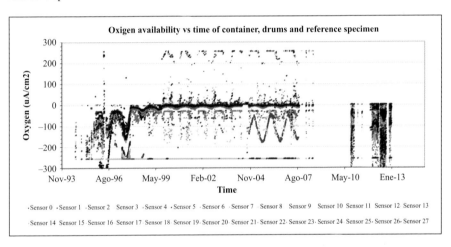

Figure 8.12 Evolution with time of oxygen availability in the concrete, mortar and reference specimen.

being produced. The trend in the oxygen can then be understood as an increase in the resistivity, which can indicate a continuous decrease in the amount of liquid water in the pores, due probably to progressive hydration. The lack of electrolyte would justify the values of almost no or very little oxygen availability. This hypothesis has to be confirmed in the future, as from 2007, when the new instrument was installed, larger values of the cathodic current were recorded.

As indicated before, the corrosion potentials (Figure 8.13) barely change (except for the values of the sensors that have failed), although a slow progression towards more positive values is clear. What is even more interesting is verification that the corrosion potential is different (more positive for ~100–150 mV) in the reinforcement embedded in the container than in the steel of the drum embedded in the filling mortar. From 2007, due to installation of the new instruments, some variations in the previous

196

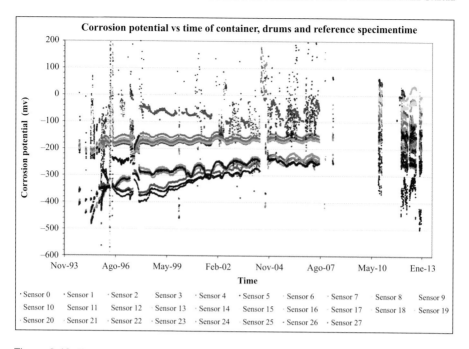

Figure 8.13 Evolution with time of corrosion potential in the concrete, mortar and reference specimen.

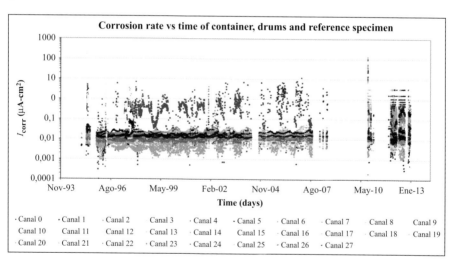

Figure 8.14 Evolution with time of the corrosion rate in the concrete, mortar and reference specimen.

values recorded were observed with the values of corrosion potential for the steel of the drums always being more cathodic than those of the reinforcement of the container.

With respect to the corrosion rate (Figure 8.14), until 2007 only the sensors that had failed presented values above 0.1 µA/cm². The others showed perfect passivity of the steel, showing, logically, the seasonal variations in temperature. From 2007 some sensors exhibited corrosion rate values above 0.1 μA/cm² in parallel with more cathodic corrosion potentials. It seems premature for a depassivation to have occurred. Further recordings will allow an understanding of whether it is a problem with the new instruments or represents a real depassivation trend.

Finally, regarding resistivity, Figures 8.9 and 8.15 show increases in time and that the effect of the seasonal evolution of temperature is greater as time progresses.

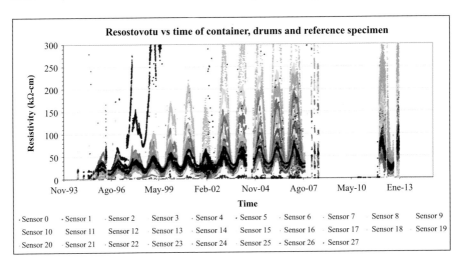

Figure 8.15 Evolution with time of the resistivity in the concrete, mortar and reference specimen.

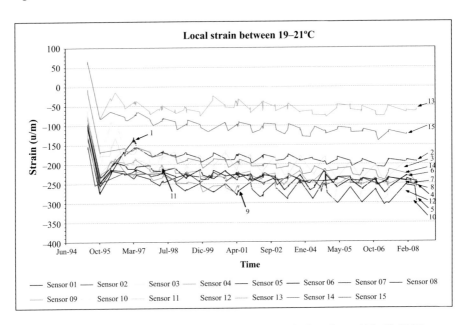

Figure 8.16 Evolution of local strain with time by filtering only the values within 19–21°C range.

Looking at Figure 8.9 it can be deduced that the highest seasonal variations are presented by sensors located in the lowest positions. This has to be analysed with the previous appreciation about oxygen availability during progressive hydration, and would imply that a drier concrete will show more pronounced resistivity changes due to temperature than a wetter concrete. The increase with time of the resistivity is better appreciated in Figure 8.16, where the evolution of resistivity values at $20 \pm 2°C$ is plotted. This graph reveals that at the beginning the evolution is exponential, but a straight line can be plotted later.

8.5 Discussion

The two main features commented upon with the recorded results are as follows:

- The effect of temperature on the different parameters under buried conditions
- The possibility to predict performance in the long term

As mentioned, temperature is a key parameter, which decisively influences everything else. Seasonal differences are crucial. It was also noticed that buried depth has less impact. As the corrosion potential and corrosion rate were found to vary less due to the steel passivity, microstrains and resistivity with respect to temperature will be analysed in more detail in the future.

- With respect to the local strain sensors embedded in the container walls in the concrete and in the mortar filling, the analysis consisted of averaging values from the sensors and separating those from the container and the mortar filling. Some outlier values were discarded from analysis.

  The main shrinkage of ~200 µm was detected in the initial years, and the later shrinkage was considered less important. For the reference specimen located in the side receptacle the averaged local strain varied seasonally from 25 µm to between −125 and −200 µm. The buried depth showed no significant differences in the drums.

  With respect to the differences between the concrete and the mortar filling, the mortar had slightly more shrinkage than the concrete. Thus, while the maximum local strains in the concrete were around 200 µm, the drums (mortar filling) showed those values only in the upper levels and in the lower levels the local strains were less than 150 µm.

  For simplicity, further analysis of the results was carried out by applying Arrhenius Law (where E_a is the 'apparent activation energy', in J/mol; R is the constant of gases and T is the temperature in K),

$$Ln\delta = Ln\delta_0 - \frac{E_a}{R \cdot T} \tag{8.3}$$

  to the trends of the average values of strain values between 19 and 21°C (Figure 8.16). Table 8.1 gives the values of the activation energy and the ordinate in the origin of the container and in the mortar filling (in the drums), together with the unitary variation of local strain by temperature after the

Table 8.1 Values of apparent activation energy of local strain variation within temperatures of 19–21°C.

Local strain	Apparent activation energy, E_a (J/mol)	Ordinate in the origin, $Ln\delta_0$	Unitary seasonal variation of local strain (µm/°C)
Concrete in the container	9,638.71	1.61	2.652
Mortar filling around drums	21,712.75	−3.95	3.901

Table 8.2 Average values of the variation of local strain with time.

Local strain	Relation between local strain and time t	Prediction to 50 years	Prediction to 300 years
Concrete	$\delta_{20} = -4.603t - 204.048$	−434.198 µm	−1,584.948 µm
Mortar	$\delta_{20} = -9.372t - 124.948$	**−593.548 µm**	−2,936.028 µm

initial shrinkage has occurred. The unitary variation is due mainly to seasonal temperature variations.

Discarding the initial shrinkage enabled further analysis of the variations of local strains due to seasonal temperature variations by fitting a straight line to the trends in Figure 8.16. The resulting expressions are giving in Table 8.2. The values obtained enabled it to be determined that the concrete shrinks at a rate of −4.603 µm/year and the mortar shrinks −9.377 µm/year.

- Regarding resistivity, its representation versus temperature is presented in Figure 8.17. From this figure, it can be noted that at 20°C a significant change in the negative slopes occurs. Above 20°C the slopes change little, although they evolve marginally because of the increase in resistivity with age. However, below 20°C the slopes change significantly with time. It can be deduced that as the concrete dries, and hydration changes with time, the changes in resistivity are more pronounced the lower the temperature.

From Figure 8.17 an apparent activation energy can be calculated from $R = R_0 e^{-Ea/RT}$ (Andrade *et al.*, 2011). The values in the concrete varied between 27,715.53 J/mol and 124,710.07 J/mol while in the mortar the values were between 25,523.41 J/mol and 88,293.60 J/mol. The average apparent E_a values of the sensors embedded in the container show higher values than those embedded in the drums and are higher than the typical values of electrolytes (20,000 J/mol). Regarding the comparison with values found by other authors (Schiessl and Raupach, 1994; Brameshuber, 2009), those of the concrete are higher (Table 8.3). Only the values found in the drums present similar levels.

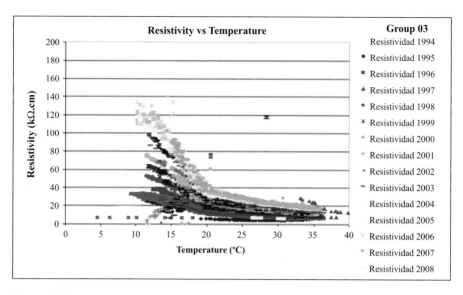

Figure 8.17 Resistivity values versus temperature, allowing calculation of the apparent activation energy.

Table 8.3 Values of apparent activation energy of resistivity variation with temperature.

Activation energy of resistivity	Average E_a > 20°C (J/mol)	Average E_a < 20°C (J/mol)	Average E_a value (J/mol)
Concrete	61,485.9	62,181.8	61,833.8
Mortar	39,726.3	39,732.9	39,729.6
Brameshuber		24,943.42–45,729.60	
Raupach		17,709.83–45,729.60	

It can be seen that the apparent activation energy does not exhibit a unique value, but varies considerably. Consequently it is difficult to select a single value to be used for long-term prediction. It would appear to be more appropriate to select values in the range of 19–21°C, as plotted in Figure 8.18, and fit into these values the trend (shown in Table 8.4).

- Finally with respect to the oxygen availability, the flow of oxygen for a cathodic current of 10 µA/cm² would be

$$Flow_{O2} = \frac{10 \times 10^{-6}}{4 \times 96,484.56} = 2,591 \times 10^{-11} mol/cm^{-2}/s^{-1} \qquad (8.4)$$

For long-term prediction it can be summarised that the results recorded provide valuable information that can be incorporated into models for performance prediction, but it must be emphasised that none of the existing models for the prediction of corrosion

201

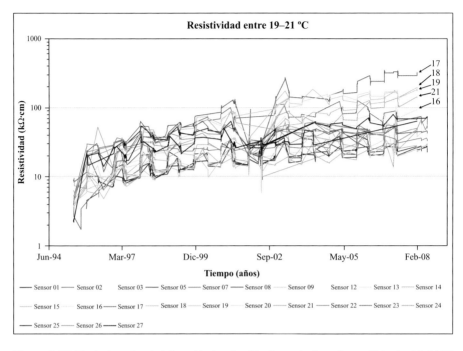

Figure 8.18 Evolution of resistivity with time by filtering only the values within the 19–21°C range.

Table 8.4	Average values of the variation of resistivity with time.	
Resistivity	Relation between local strain and time t	Prediction to 50 years kΩ cm
Concrete	$\rho_{20} = 10.30e^{0.1623t}$	34,459
Mortar	$\rho_{20} = 22.62e^{0.078t}$	**1,117.43**

of the reinforcement have been validated for a sufficiently long period of time. However, for the damage scenarios considered, the sensors are not recording the advance of the carbonation front. Only the corrosion onset, if produced, could give information on the carbonation rate that would be produced in buried conditions. The sensors would then be reporting information on the 'aging' of the concrete and on the oxygen availability. At present, a model for this aging does not exist, and this would form the basis of an interesting research project.

8.6 Conclusions

The conclusions that can be drawn are as follows:

- The percentage of failed sensors was less than 18% of the total, and occurred mainly during container fabrication. From installation of the new Geologger some of the sensors produced outlier values for oxygen content, corrosion potential and corrosion rate that need further observation to be interpreted correctly.

- Temperature is the main controlling parameter, due to seasonal variations. The temperature varies between 5 and 35°C.
- In the concrete and in the mortar there was initial shrinkage. This was attenuated later and appeared to stabilise after 20 years. The local strains were quantified with an apparent activation energy and modelled its evolution with time after the first initial change.
- The resistivity is increasing steadily with time and appears not to stabilise. The resistivity evolution is attributed to continuous hydration or concrete 'aging'. Apparent activation energies were quantified and show interesting scatter and variations. Also, the trend of resistivity with time has been modelled as an exponential expression.
- The oxygen availability is decreasing with time, which has been interpreted as being due to the progressive disappearance of pore water due to continuous hydration. Some sensors again gave high values after installation of the new Geologger, which need further monitoring and investigation.
- The corrosion potential and corrosion rate show values of passivity. Some sensors gave outlier values after installation of the new Geologger. These need further investigation.

Acknowledgements

The authors acknowledge funding provided by Enresa-Spain to support this investigation.

References

Andrade, C. and J.A. Gónzalez (1978) Quantitative measurements of corrosion rate of reinforcing steels embedded in concrete using polarization resistance measurements. *Werkstoffe und Korrosion* **29**, 515.

Andrade, C., J. Rodriguez, F. Jimenez, J. Palacio and P. Zuloaga (2000) *Embedded sensors for concrete structures instrumentation*. OECD-NEA Workshop on Instrumentation and Monitoring of Concrete Structures, Brussels, Belgium, pp. 249–256.

Andrade, C., I. Martínez, M. Castellote and P. Zuloaga (2006) Some principles of service life calculation of reinforcements and in situ corrosion monitoring by sensors in the radioactive waste containers of El Cabril disposal (Spain). *Journal of Nuclear Materials* **358**, 82–95.

Andrade, C., P. Zuloaga, I. Martinez, A. Castillo and S. Briz (2011) Effect of temperature on the corrosion parameters and apparent activation energy measured by embedded sensors in a pilot container in 'el Cabril' repository. *Corrosion Engineering, Science and Technology* **46**(2), 182–189.

Brameshuber, W. (2009) Monitoring of the water distribution in concrete structures. In *2nd International RILEM Workshop on concrete durability and service life planning* (ConcreteLife 09), 7–9 September, Haifa, Israel, pp. 146–154.

Feliú, S., J.A. González, S. Feliú Jr and C. Andrade (1990) Confinement of the electrical signal or in-situ measurement of polarization resistance in reinforced concrete. *ACI Material Journal* **87**, 457.

Feliú, S., C. Andrade, J.A. González and C. Alonso (1996) A new method for in-situ measurements of electrical resistivity of reinforced concrete. *Materials and Structures* (Rilem) **29**, 362–365.

Gjørv, O.E. and Ø. (1983) Vennesland Diffusion of dissolved oxygen through concrete. *Materials Performance* **25**, 39–44.

Polder, R., C. Andrade, B. Elsener, O. Vennesland, J. Gulikers, R. Weidert and M. Raupach (2000) Test methods for on site measurement of resistivity of concrete, RILEM TC 154-EMC: Electrochemical techniques for measuring metallic corrosion. *Materials and Structures* (Rilem) **33**, 603–611.

Rilem Recommendations of TC-154-EMC (2004) Electrochemical techniques for measuring corrosion in concrete – test methods for on-site corrosion rate measurement of steel reinforcement in concrete by means of the polarization resistance method. *Materials and Structures* **37**, 623–643.

Schiessl, P. and M. Raupach (1994) Influence of temperature on the corrosion rate of steel in concrete containing chlorides. In *First International Conference of Reinforced Concrete Materials in Hot Climates*, April 1994, United Arab Emirates University, El-Ain, UAE, pp. 537–549.

Zuloaga, P. (2010) *Spent nuclear fuel management in Spain*. IAEA International Conference on Spent Fuel from Nuclear Power Reactors, 31 May–4 June 2010, Vienna, Austria, pp. 12–17.

Chapter 9
Reactions of cements in geothermal wells

Neil B Milestone

The standard practice for cementing geothermal wells is to use a slurry containing 40% silica flour together with an API Class G or H cement. This reduces the Ca/Si ratio of the binder and allows the formation of 11 Å tobermorite to counter the phenomenon of 'strength regression', where strength slowly decreases with time due to crystallisation of the high Ca/Si ratio phase known as a-C_2SH. However, while tobermorite provides an excellent binder, along with other low Ca/Si ratio phases, it undergoes rapid carbonation from any dissolved CO_2. If the concentration of CO_2 is high enough, the geothermal fluids are mildly acidic due to 'carbonic acid' and will dissolve the calcium carbonate formed, leading to corrosion and loss of calcium as soluble calcium bicarbonate. Problems have been encountered in several New Zealand fields where the cement has corroded so the steel casing is no longer protected and it also corrodes, leading to casing failure.

This chapter describes some of the reactions that occur on hydrothermal treatment of Portland cements, and details ways of improving cement to counter this CO_2-induced carbonation and subsequent corrosion.

9.1 Introduction

Geothermal steam has been used for centuries in many applications throughout the world. Heat, or rather the steam supplied from the centre of the Earth, provides one of the more successful ways of obtaining renewable energy, provided it can be suitably harnessed. Up to the end of 2009, 423,483 TJ of energy was being extracted annually through direct use installations (Lund *et al.*, 2010). While very large amounts are used for heating, either through heat pumps, space and district heating, or in agriculture, efforts have concentrated on electricity production, with installed power of over 11,000 MWh (Salim and Amani, 2013). New Zealand generates around 15% of its electricity from geothermal steam.

9.1.1 Cements for geothermal wells

To extract steam from geothermal fields, wells are drilled into a formation, which is usually weak and often porous. A permeable formation is necessary to ensure a good supply of steam. Drilling of the well is carried out in stages with different-diameter holes into which a series of steel casings are lowered and cemented in place using a cement slurry pumped through a hollow drill stem to return to surface through the annulus between the casing and formation, providing a cement sheath around the outside of the steel casing and ensuring it is anchored to the formation. In a completed well there are usually three concentric casings, each with its own particular function and cement annulus: the *surface casing* (typically 23-inch outer diameter (OD)) for the first 20 m or so to consolidate the formation and provide a base for further drilling, an *anchor casing* (typically 13⅜-inch OD) for the next 100 m, which as its name implies ensures the well remains in the ground, and the *production casing* (typically 7⅝-inch OD), which runs the length of the well to prevent well collapse and through which steam is delivered. The main valve sits atop this casing. The production zone is usually tapped with a per-forated, uncemented liner hung from the bottom of the production casing (Figure 9.1).

The cement or grout sheath around the steel casing has several important roles to play in a well:

- It provides a seal that isolates the well bore from the formation fluids so there is no mixing of fluids or pressures from the zones encountered up and down the casing.

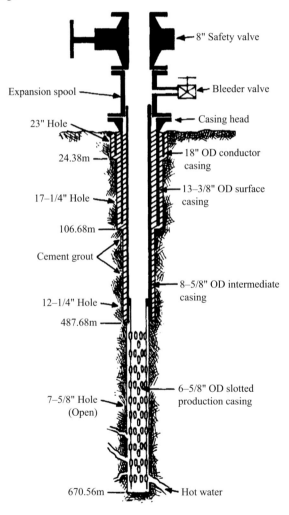

Figure 9.1 Typical section of well casings.

- It ensures the high pressures encountered down hole are contained within the casing, allowing a 'shut off' of the well and avoiding a 'blowout'.
- It provides alkaline protection to a steel casing, preventing corrosion.
- It helps support the steel casing against fatigue due to vibration etc. and ground movement.

The steel casings have traditionally been cemented to the rock formation in both oil and geothermal wells using cement slurry formulations based on calcium silicate cements, typified by Portland cement. Portland cement is a calcium silicate cement made by firing limestone and clay to high temperatures. The two main phases that make up over 80% of the material are tricalcium silicate (C_3S) and dicalcium silicate (β-C_2S). When Portland cement reacts with water, it hydrates to form two main compounds, crystalline calcium hydroxide, $Ca(OH)_2$, which provides the high pH, and an amorphous binder known as calcium silicate hydrate gel, C-S-H (cement oxide nomenclature: $C = CaO$, $S = SiO_2$, $A = Al_2O_3$, $F = Fe_2O_3$, $H = H_2O$, $N = Na_2O$). While Portland cements have been used extensively in construction concrete for over 150 years, the physical requirements of the slurry for well cements are very different to those for construction cements, although essentially the same phases form on hydration at ambient conditions. Both types of cement rely principally on the hydration of the two calcium silicates, C_3S and β-C_2S, to form a calcium silicate hydrate binder, normally designated C-S-H. The composition of this binder is not fixed but varies during hydration as well as with the temperature of hydration. At ambient temperatures, C-S-H has an approximate formula of '$C_3S_2H_3$', giving a Ca/Si ratio typically around 1.5. In the early stages of hydration, this ratio can be higher, decreasing as hydration proceeds. Significant quantities of $Ca(OH)_2$ are also formed as the calcium silicates react with water or hydrate. However, when Portland cements hydrate in hydrothermal conditions, the calcium silicate hydrate crystallises to a series of phases, the exact composition depending on the overall Ca/Si ratio of the binder and temperature, as discussed in Section 9.2.

9.1.2 API specifications

The cements used in wells throughout the world are now almost universally designated by the American Petroleum Institute (API) Class codes (A–H). Some of the differences between these cements have been described by Bensted (1989). The most common classes found are A, G and H, with G and H existing in two forms, moderate and high sulfate resistance. A Class J cement was specified in earlier classifications but is no longer readily available. It was not a Portland cement but rather a β-C_2S plus quartz blend, specifically designed for high-temperature applications.

Many of the physical properties of well cements are based on API imperial measurements, giving units such as lbs/US gal for density and US gals/sack (94 lbs) for additives. Casing dimensions are in inches.

Other non-API cements have been prepared specifically for high-temperature applications, such as Geoterm® (Italcementi), which is an interground low C_3A Portland clinker/quartz/gypsum cement, and ThermaLock®, a non-Portland aluminate/phosphate based cement from Halliburton. New cements, such as EverCrete® from Schlumberger and CorrosaCem® from Halliburton, have been designed for the CO_2

storage market and also appear to be non-Portland cements and may be applicable for geothermal applications, but there is limited information available on their properties.

9.2 Effects of temperature on cement hydration

It is well known in construction that if precast concrete samples are 'steam cured' above 60°C at atmospheric pressure, there can be a reduction in strength. Paul and Glasser (2000) attributed this to a denser hydration product that formed quickly and sealed the surface of the cement particles, preventing ongoing hydration. Menzel (1934) found that when a cementitious composite was autoclaved, there was a notable reduction in strength over a conventionally cured composite. However, it was work by Kalousek and Adams (1951) that showed that this drop in strength was due to crystallisation of the calcium silicate hydrate to give a well-defined product known as alpha dicalcium silicate hydrate (α-C_2SH) along with calcium hydroxide ($Ca(OH)_2$). The crystals of α-C_2SH are dense and contain little water, so the matrix becomes permeable and is weak. In the highly alkaline matrix dictated by the presence of $Ca(OH)_2$, the α-C_2SH crystals continue to grow, giving rise to the phenomenon known as *strength retrogression*, where the compressive strength slowly decreases with time and the matrix becomes porous.

When hardened cement paste is heated in air, dehydration occurs, with loss of water from the C-S-H between 80 and 140°C and a slow ongoing loss until the $Ca(OH)_2$ decomposes around 440°C and all strength is lost.

9.2.1 Hydrothermal cement curing

Under hydrothermal conditions, dehydration does not take place and a series of crystalline hydrates are formed, the compositions of which depend on the available Ca/Si ratio, the type and amount of any additional silica present, and the temperature. Around 150°C, α-C_2SH crystallises from pure Portland cement along with $Ca(OH)_2$, while at higher temperatures, a mixture of phases can form: reihardbraunsite (C_5S_2H), jaffeite ($C_6S_2H_3$) and kilchoanite (C_3S_2) (Milestone *et al.*, 1992; Bresson *et al.*, 2002; Meducin *et al.*, 2008). These phases all coexist with $Ca(OH)_2$. However, these high Ca/Si ratio phases crystallising from pure Portland cement are not good binders and continue to lose strength, becoming more porous and permeable, and measures are usually taken to avoid their formation in high-temperature well cementing.

Kalousek (1954, 1955) found, by adding >40% by weight of cement (BWOC) of silica flour (fine quartz), that strength retrogression could be avoided with the formation of a further set of crystalline phases. At 150°C he found quartz reacted to give a low Ca/Si ratio phase, 11 Å tobermorite ($C_5S_6H_5$). Tobermorite crystallises in a fibre-like form that includes water in its structure, so it is a lot less dense than α-C_2SH. The fine needle-like crystals lock together well, so tobermorite functions as a very good binder. Any $Ca(OH)_2$ from hydrating cement reacts with the quartz, forming additional binder, so the composites are strong and impermeable. Additions of around 40% quartz flour have become the norm for wells where down-hole temperatures exceed ~120°C, and this avoids the problems associated with strength retrogression.

Above 200°C, tobermorite converts to a second low Ca/Si ratio phase, xonotlite (C_6S_6H). While xonotlite is also fibrous and retains much of the strength of tobermorite, it contains less water and is slightly more permeable, although the strength remains comparable. In recent work, Iverson *et al.* (2010) have suggested that even adding 40% silica flour may not ensure that strength is maintained and they have recommended that up to 60% be added.

Tobermorite is the binder phase in autoclaved cement/fibre products. These are manufactured from a slurry of ~40–60% quartz flour, cement and ~8% cellulose fibre via the Hatscheck process. The green product is allowed to hydrate, often with steam curing, for several hours, before undergoing autoclaving at 175°C for up to 16 hours. However, in this short curing time, only the finer fraction of quartz reacts, and a layer of tobermorite forms around the quartz grains, which acts as a form of fine reinforcement.

The reaction of $Ca(OH)_2$ with fine quartz is slow below 150°C. It was found that additions of aluminium-containing kaolinite or metakaolinite greatly improved the rate at which quartz reacted (Klimesch, 1999), and also that the reaction usually proceeded via α-C_2SH with the tobermorite product formed containing aluminium with some advantageous properties.

Xonotlite blocks made from a lime/quartz slurry are used in insulation and as the porous support for acetone in acetylene cylinders. Above 200°C, where additional quartz flour has been added, xonotlite (C_6S_6H; Ca/Si = 1) usually forms in geothermal cements, although it will form at temperatures below this, particularly if additional additives are included (Milestone and Ahari, 2007).

9.2.2 Other crystalline calcium silicate hydrates

Various calcium silicate hydrates can form within the $CaO/SiO_2/H_2O$ phase diagram and these have been extensively addressed in early papers by Heller and Taylor (1951, 1952a,b), Peppler (1955) and Assarsson (1956, 1957,1958, 1960), and later reviews by Taylor (1968), Taylor and Roy (1980) and Richardson (2008). In most of these studies, the phases have usually been prepared from a $CaO(Ca(OH)_2)/SiO_2/H_2O$ system rather than from Portland cement. A number of the phases have only been identified from rare, naturally occurring crystals, as the pure phases have yet to be synthesised. In geothermal wells the starting material is invariably a Portland cement blend, and the phases that form from this combination are more limited and are discussed below.

Apart from tobermorite and xonotlite, phases such as gyrolite ($C_2S_3H_3$; Ca/Si = 0.66) or trusscotite ($C_7S_{12}H_2 nH_2O$; Ca/Si = 0.58) can sometimes form with higher silica loadings, although we have found these phases rare in the geothermal cements we have studied and special conditions tend to be needed to allow these phases to form (e.g Nelson and Kalousek, 1977; Eilers *et al.*, 1983; Nelson *et al.*, 1991). The presence of small amounts of CO_2 or $CaCO_3$ can give rise to scawtite ($C_7S_6(CO_3)H_2$; Ca/Si = 1.17), while addition of sodium silicate gives rise to pectolite (NC_2S_2H; Ca/Si = 1.0) (Eilers *et al.*, 1983). These low Ca/Si ratio phases are all good binders with low porosity.

Even non-optimum additions of silica to calcium-silicate-based cements have marked effects on both strength and permeability. Still further new phases can be formed depending on the Ca/Si ratio and temperature, such as killalaite (C_6S_4H; Ca/Si = 1.5) (Milestone *et al.*, 1992), foshagite (C_4S_3H; Ca/Si = 1.33) and hillebrandite ($C_2S_3H_2$; Ca/Si = 0.66) (Yong and Glasser, 2004).

As mentioned above, the quartz/$Ca(OH)_2$ reaction is rather slow and the use of other forms of silica has been examined. However, these other forms of silica tend not to give rise to tobermorite and so their use has not been actively pursued, despite their better carbonation resistance, as discussed in Section 9.4. Why only quartz gives rise to tobermorite is not well understood, but is believed to be associated with the rate of solubility of the quartz (Jupe *et al.*, 2008; Kikuma *et al.*, 2011).

9.3 Durability

A key factor in establishing the life of geothermal wells is how complete the cement sheath is and how long it will last in what is a very harsh environment. In some wells the fluids can be very acidic, with pH as low as 2.5, as found in the Philippines (Brondial *et al.*, 2000; Nogara *et al.* 2002; Brondial, 2005), while temperatures often exceed 300°C (Quinao *et al.*, 2013). While the low pH in the Philippine fields is caused by acid sulfate waters, in New Zealand the pH is not as low, typically reaching only 4.7 due to large amounts of dissolved CO_2. In any acidic environment, issues with corrosion of both cement and steel are of concern.

9.3.1 Preliminary examinations of cements exposed downhole

The wells drilled in the 1950s for the new Wairakei geothermal power station in New Zealand were cemented using plain Portland cement with addition of a few percent of a local Porongahau bentonite to provide a suitable slurry density. Generally, few problems have been encountered in the well cementing at Wairakei, although Kennerley (1961) showed that for samples of the cement recovered from wells in that field, many had skins of calcite and α-C_2SH was the phase present as expected. Carbonation was limited to a thin skin.

When cementing was considered for the new wells to be drilled for the Ohaaki power station in the 1980s, recommendations based on the work by Kalousek (1955) were to use a Portland/silica flour blend with the new API class of cements, as strength retrogression was uppermost in the minds of the drilling engineers. A series of formulations were prepared and exposed in well BR23. Results from the downhole exposures after 6 months were surprising. Extensive carbonation had occurred and many samples had suffered severe corrosion (Milestone *et al.*, 1986a). While some samples showed a skin of calcite, the expected calcium carbonate phase, other samples had almost completely converted to aragonite with up to 60% of the volume of some samples lost by corrosion.

Workover on wells designated for production, which had been drilled several years before and kept shut in, showed that for several wells, the casings had corroded and perforated (Driver and Wilson, 1984; Bixley and Wilson, 1985). This fact, coupled with the initial results from the cement exposure, prompted a re-evaluation of the field chemistry and resulted in a new picture of the geothermal fluids. Hedenquist (1990) identified a low-temperature aquifer in which CO_2, boiled off from near the production zone, had percolated through the porous formation and become trapped beneath an impermeable capping layer, which lay several hundred metres below surface. The high levels of dissolved CO_2 within this trapped aquifer rendered the fluid mildly acidic with 'carbonic acid'. The temperature of these 'corrosive zones' is typically around 150–180°C, much lower than the production zone. Similar zones have since been identified in other fields where a cooler layer of water is heated by uprisings from deeper fluids and corrosive gases are dissolved (Clearwater *et al.*, 2012).

9.3.2 Carbonation and corrosion in geothermal wells

Milestone *et al.* (1986a) showed that samples that had been precured in a laboratory autoclave at 150°C for 72 hours, and then exposed downhole in a well with high CO_2-containing fluids in the Ohaaki geothermal field, underwent both extensive carbonation and corrosion at 150°C (Figure 9.2). The outside layer of these samples was a slippery coating of silica gel from which all calcium had been leached. At 260°C, only carbonation had occurred when they were exposed to the high-temperature production fluid (Figure 9.3).

In laboratory work carried out at Brookhaven National Laboratory by Milestone *et al.* (1986b,c, 1987a,b), together with field exposure data, it was revealed that when 40% silica flour was added to counter strength retrogression, a matrix was created that rapidly carbonated. Carbonation of the low Ca/Si ratio phases of tobermorite and xonotlite was rapid and they became porous. In contrast, the low-strength, high-Ca/Si-ratio phases of α-C_2SH (at 150°C) and reinhardbraunsite and kilchoanite (at 260°C), which crystallised in association with $Ca(OH)_2$, carbonated to produce a dense, protective carbonation sheath with low permeability and high strength that limited carbonation to a surface rim only a few millimetres deep.

$Ca(OH)_2$ carbonates to calcite, whereas at 150°C, calcium silicate hydrates carbonate to metastable aragonite, so in the samples that also contained $Ca(OH)_2$, mixtures of aragonite and calcite were present in the thin carbonated layers. At 260°C, both $Ca(OH)_2$ and the calcium silicate hydrates form calcite upon carbonation. Thus, when tobermorite carbonates, it forms aragonite at 150°C while xonotlite forms calcite.

(a)   (b)

Figure 9.2 Exposure of grout cylinders in BR23 Ohaaki (150°C, $[CO_2] = 0.35$ moles/l, pH_T 4.59) for 6 months. (a) Pure cement: The centre core is α-C_2SH plus $Ca(OH)_2$, the middle orange layer is calcite, with an inner layer of calcite and aragonite, and the outermost layer is silica gel. Minor corrosion has occurred. (b) Cement plus 40% silica flour: The inner core is aragonite, the middle layer is aragonite/calcite and the outer layers are corroded silica gel. The sample is extensively corroded.

(a)   (b)

Figure 9.3 Exposure of grout cylinders in BR17, Ohaaki (260°C, [CO_2] = 0.064 moles/l, pH_T 6.49) for 6 months. (a) Pure cement: The inner core is high-Ca/Si phases kilchoanite and reinhardbraunsite and the outer rim is calcite. No corrosion. (b) Cement plus 40% silica flour: The inner core is xonotlite and calcite, the outer core is calcite and xonotlite and the outside layer is calcite. No corrosion.

Following his work at Brookhaven Laboratory, Milestone and co-workers in New Zealand (Milestone and Aldridge, 1990) carried out an extensive period of downhole testing in both the Ohaaki and Rotokawa fields, using a wide variety of formulations based largely on additions to Portland cement. It became clear from these studies that in any Portland cement-based system, the carbonation rate was very dependent on the volume of available Ca in any unit volume and particularly any $Ca(OH)_2$ present. Carbonation resistance was enhanced by low addition rates of silica and low water/solids ratio mixes. Addition of bentonite, often added to control bleeding or segregation, proved detrimental, as it stabilised tobermorite formation, which rapidly carbonated and became porous (Milestone *et al.*, 1987b). It was also found that as the temperature increased, a bentonite-stabilised slurry became unstable and segregated.

9.4 Mechanism of carbonation

Carbonation of cement grouts proceeds via a 'through solution' mechanism, as initially described by Milestone *et al.* (1986a). Ca^{2+} ions migrate through a saturated matrix, leaving a leached zone, and then rapidly precipitate as $CaCO_3$ when they meet the dissolved CO_2 ('carbonic acid'). This reaction can lead to a discontinuity as the precipitation reaction occurs faster than the rate at which the Ca^{2+}ions can migrate. The discontinuity is obvious in Figure 9.4, where a sample of pure cement exposed in an Ohaaki well (BR17, [CO_2] = 0.064 moles/kg) at 260°C for 4 months shows the formation of a calcite skin that is only weakly attached and will readily separate, particularly if the sample is recovered from the well too quickly and dried out. The pH_T in this well is 6.49, significantly above the neutral pH_T of 5.82, so there is no acid dissolution or corrosion.

Figure 9.4 Exposure of pure cement samples in BR17 at 260°C for 4 months.

Figure 9.5 Photomicrograph of carbonated geothermal cement sample from Figure 9.3. Scale bar is 1 mm. Brown colour is calcite, white crystals are αC_2SH and grey is $Ca(OH)_2$. The discontinuity is ~0.7 mm and the leached zone ~1.2 mm.

Carbonation of $Ca(OH)_2$ is an expansive reaction, so the carbonated layer occupies more volume than that from where it has come and can separate. A petrographic thin section prepared from the same sample (Figure 9.5) clearly shows the leached region along with the discontinuity.

Ca^{2+} ions migrate from the sparingly soluble $Ca(OH)_2$, leaving a leached zone in which the loss of $Ca(OH)_2$ is clearly visible just inside the discontinuity. Ultimately, the skin of $CaCO_3$ will grow and penetrate to the centre with an increase in the volume of the cylinder, but after 1 year of downhole exposure, the carbonated layer was only around 12 mm thick.

Formation of the carbonation layers has important issues if compressive strengths or permeabilities of cylinders are being used as measures to determine potential

213

durability. The carbonate skin has high strength, so what is measured for strength is due to a composite structure that may contain a discontinuity. The same applies when measuring permeability, where water is likely to flow down the discontinuity rather than through the sample. Carbonation per se does not lead to decreased strengths; rather it enhances it.

The calcium silicate hydrates are much less soluble than $Ca(OH)_2$ meaning there are limited Ca^{2+} ions to migrate so the carbonation attack appears to occur directly on the matrix, as seen in Figure 9.2b, where a sample containing 40% silica flour exposed in the same well has no obvious leached zone. The carbonation of xonotlite, the stable low Ca/Si ratio phase at 260°C, also results in a reduction of solid volume, causing an increase in porosity, so no distinct carbonation layer forms.

9.4.1 Available Ca^{2+} per unit volume

As mentioned above, the concentration of Ca per unit volume has a marked effect on the rate of carbonation. One of the issues in cementing a geothermal well is that it is usually drilled through a weak formation, often with zones where lost circulation is a problem. As a consequence, slurries with low density are sought to ensure the weight of the column of cement slurry does not collapse the formation. This was originally addressed in New Zealand by using formulations with higher water/cement ratios with addition of bentonite, which was shown by Milestone *et al.* (1987b) to cause greater amounts of carbonation, while the slurries themselves were not stable at elevated temperatures. With silica flour additions, bentonite enhanced and stabilised the formation of tobermorite, which is not carbonation resistant. The use of hollow ceramic cenospheres or microspheres to obtain lightweight slurries is now widespread. These are obtained from fly ash and have the same overall composition so are pozzolanic. They allow slurry densities as low as 1.55 g/cc (13.5 ppg) to be easily obtained.

In a recent study, a slurry of Class G plus 20% amorphous silica BWOC (described in Section 9.6.2) and 15% of cenospheres BWOC (14 lb/sack) was mixed at a water/solids ratio of 0.41 to give a density of 1.65 g/cc (13.8 ppg), then cured at 165°C for 28 days in a simulated carbonated geothermal brine (15.8 g NaCl, 4.1 g KCl, 0.2 g Na_2SO_4, 0.05 g $CaCl_2$, 15.6 g CB8 colloidal silica (50% solids) dissolved in 20 l distilled water) under 6 bar CO_2 pressure. While compressive strength was maintained (25.7 MPa), there was a distinct carbonation rim of ~5 mm around the sample that increased with time. A photo-micrograph of the core (Figure 9.6) shows distinct reaction rims around the cenospheres, many of which are filled with aragonite and calcite. An X-ray of the external carbonation rim showed it to be totally carbonated with largely calcite present, whereas in the core, tobermorite and a trace of xonotlite were the crystalline calcium silicate hydrates present, with large amounts of aragonite and a smaller amount of calcite.

The amorphous silica reacts quickly with any $Ca(OH)_2$ produced from cement hydration, effectively removing it. The cenospheres are also siliceous and readily react with Portland cement hydration products to give pockets of low Ca/Si ratio phases that crystallise as tobermorite or xonotlite but that carbonate readily. In a Class G plus 25% amorphous silica slurry where cenospheres were not added, the density was 1.89 g/cc (16.5 ppg) and there was no carbonation beyond the first few millimetres. Thus, in the lightened samples where the amount of Ca per unit volume was low, carbonation was rapid as no protective sheath was formed.

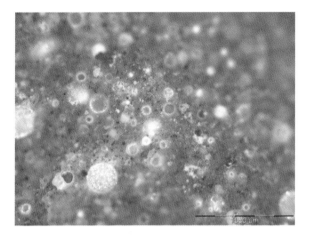

Figure 9.6 Photomicrograph of Class G cement plus 25% hydrothermal silica and 15% cenospheres mixed at a water/solids ratio of 0.41 and cured for 28 days in CO_2 saturated brine at 160°C.

(a)　　　　　　(b)　　　　　　(c)

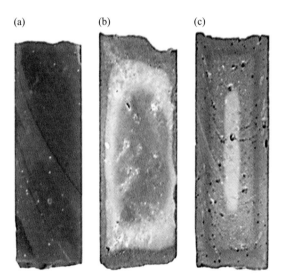

Figure 9.7 Grout formulations containing 20% silica at different densities cured at 150°C: (a) 15.0 ppg, (b) 12.4 ppg and (c) 11 ppg.

This effect is further highlighted in a second study (Milestone *et al.*, 2014), where the slurry density of a Class G cement plus 20% hydrothermal silica was controlled with water addition and additives and the samples were cured under the same high CO_2 conditions. The lighter weight slurries had carbonated through to the core (Figure 9.7).

The binder in the Class G plus 20% amorphous silica slurry with a density of 16.5 ppg discussed above was almost completely amorphous after 28 days curing, with only a very thin carbonation rim. In the 15 ppg sample, the major phase was an amorphous C-S-H with small amounts of xonotlite and afwillite ($C_3S_2H_3$), and with traces of calcite along with a significant carbonation rim. As the density decreased further, the 12.4 ppg sample still contained a large amorphous component but the amounts of xonotlite and afwillite had increased while the unhydrated cement decreased and there was a small amount of gismondine, $C_2A_2S_4H_9$, present in the core. The skin contained aragonite and calcite. Finally, in the 11 ppg sample carbonation had penetrated to the centre of the sample where there was a mixture of calcite and aragonite, along with small amounts of xonotlite and the surprising presence of anhydrite. No crystalline silicates were present in the middle layer, which was a mixture of aragonite and calcite along with a small amount of anhydrite, while the skin was principally calcite, with small amounts of aragonite.

The addition of only 20% silica (particularly non-crystalline) does not normally allow the formation of the low-Ca/Si-ratio phases of tobermorite (which usually only forms with quartz) or xonotlite, as there is insufficient SiO_2 present and the Ca/Si ratio is not low enough. However, carbonation removes available Ca from the matrix and allows the formation of these lower-Ca/Si-ratio phases. Interestingly, the use of the chemical additives in these samples, which contained significant amounts of sulfate, has changed the expected range of products. The anhydrite found arises from the interaction of Ca^{2+} ions and sulfate from the additives needed to prevent settling in the higher water/solid mixes.

9.5 Mechanism of corrosion

As pointed out by Bruckdorfer (1986), the corrosion mechanism is one of dissolution of the $CaCO_3$ that has formed through carbonation of the cement (equations 9.1 and 9.2). Often, the reaction stops at this point, but if the dissolved CO_2 concentration is high enough then the 'carbonic acid' dissolves the $CaCO_3$ to give soluble $Ca(HCO_3)_2$, which is lost (equation 9.3). The samples shown in Figure 9.2 were exposed to a fluid where the dissolved CO_2 concentration (0.34 moles/l) created a mildly acidic liquid environment of pH 4.59. Neutral pH at 150°C is estimated to be 5.82 (Bandura and Lvova, 2006), so the pH is dictated by the amount of dissolved CO_2.

$$Ca(OH)_2 + CO_{2(aq)} \rightarrow CaCO_3\downarrow + H_2O \qquad (9.1)$$
$$C\text{-}S\text{-}H + CO_{2(aq)} + H_2O \rightarrow CaCO_3\downarrow + SiO_2.nH_2O \qquad (9.2)$$
$$CaCO_3 + CO_{2(aq)} + H_2O \rightarrow Ca^{2+}{}_{(aq)} + 2HCO_3{}^-{}_{(aq)} \qquad (9.3)$$

This corrosion reaction cannot occur in dry steam as there is no liquid into which the ions can dissolve and be leached, although carbonation does occur. When the CO_2 concentration is such that the fluid is not acidic, as in BR17 (Figure 9.3) then only carbonation will occur. It is difficult to simulate corrosion in the laboratory as the solution in the autoclaves saturates quickly.

9.6 The role of silica and its different forms in hydrothermal curing

As discussed above, the normal procedure in well cementing is to add 40% silica flour if the operating temperature exceeds 120°C. This allows the formation of the

Table 9.1 Chemical analysis of raw materials.			
Oxide	OPC	BFS	PFA
CaO	64.6	42.1	1.44
SiO_2	21.0	34.5	48.64
Al_2O_3	5.2	13.7	25.88
Fe_2O_3	2.6	1.0	7.57
MgO	2.1	7.3	1.42
SO_3	2.5	–	1.20
K_2O	0.6	0.5	2.83
Na_2O	0.3	0.2	1.91
LOI	0.7	−1.1	4.16

OPC: ordinary Portland cement; BFS: blastfurnace slag;
PFA: pulverised fly ash; LOI: loss on ignition.

high-strength phases of tobermorite and xonotlite and avoids strength retrogression. The reaction of quartz with $Ca(OH)_2$ is slow and carbonation can compete, removing any available Ca and reducing the amount of binder that can be formed as well as effectively lowering the Ca/Si ratio in the binder. The larger quartz particles do not react and act only as an inert filler. Thus there is the potential for using a more reactive silica as found in supplementary cement materials (SCMs) such as blast furnace slag or fly ash. These additives can react quickly with the $Ca(OH)_2$ as it forms during the initial hydration and reduce the internal pH. This could slow the growth of large crystals of α-C_2SH in geothermal wells (as seen with 20% quartz addition). The rapid reaction of $Ca(OH)_2$ with a reactive silica competes with carbonation during hydration and ensures most of the Ca becomes a calcium silicate hydrate rather than Ca carbonate. This could also assist in the preparation of autoclaved cement products where ambient curing has been shown to degrade cellulose through a combination of high pH and available Ca^{2+} ions (Milestone and Suckling, 2004).

9.6.1 Reactions of fly ash and blast furnace slag

Recently, the reaction of several types of siliceous additives and their effect on tobermorite formation and potential use for geothermal cement formulations have been examined (Milestone *et al.*, 2014). Mixtures of cement, quartz and fly ash and cement, quartz and blast furnace mixed at a water/binder ratio of 0.4 were autoclaved for up to 90 days at 160°C.

These SCMs contain significant amounts of aluminium, which becomes available as they react (Table 9.1). The pulverised fly ash (PFA) contained several percent of quartz along with mullite and hematite whereas the slag contained a small amount of gehlinite.

9.6.1.1 pH during precuring

There is a significant reduction in the pH of the initial pore solution by approximately 1 pH unit when SCMs are added (Table 9.2). The addition of quartz makes little difference to the solution pH because at ambient temperatures, no reaction occurs with

Table 9.2 pH of cement blends.

	OPC	1:1 OPC + quartz	1:1 OPC +PFA	1:1 OPC +BFS	OPC + 20% MS600
pH	13.3	13.0	12.3	12.4	12.2

OPC: ordinary Portland cement; BFS: blastfurnace slag; PFA: pulverised fly ash.

Table 9.3 Phases formed with OPC and quartz (XRD).

Composition	1 day	3 days	7 days	90 days
OPC	CH + amorphous	CH + amorphous	CH, α-C$_2$SH	CH, α-C$_2$SH
OPC : quartz (1:1)	quartz, tobermorite (tr)	quartz, tobermorite	quartz, tobermorite	tobermorite, quartz

OPC: ordinary Portland cement; XRD: X-ray diffraction.

Ca(OH)$_2$, although the slight reduction in pH is probably due to some adsorption of hydroxyl ions onto the surface of the quartz.

9.6.1.2 Phases formed upon autoclaving

OPC. Initially, Ca(OH)$_2$ was the only crystalline product formed from autoclaving OPC (Table 9.3) along with an amorphous C-S-H, but with increased curing time at temperature, α-C$_2$SH crystallised, with the amount increasing with time. After 90 days curing the samples were soft enough to be scratched easily with a pencil.

Quartz addition. With 1:1 quartz addition it took 7 days at 160°C before a well-crystallised 11 Å tobermorite was formed, although small peaks were visible at 1 day (Table 9.3). No Ca(OH)$_2$ was detected after autoclaving. After 90 days curing, while tobermorite was the predominant phase, not all the quartz had reacted.

PFA additions. With a 1:1 ratio of OPC:PFA, tobermorite did eventually form, although it took 7 days before there was any trace of it and 90 days before a significant amount was formed (Table 9.4). With higher PFA additions (OPC:PFA = 1:3), tobermorite was not formed, even after 90 days autoclaving, and the sample remained amorphous (Table 9.4). Free Ca(OH)$_2$ was not observed at any age by X-ray diffraction (XRD). When fly ash/quartz mixtures were used tobermorite did form, but the rate of reaction was slow and depended on the total amount of silica present. A typical XRD trace for the sample 1:0.5:0.5 OPC:PFA:quartz is shown in Figure 9.8. The ratio Ca/Si needs to be close to 1, with quartz present, for crystalline tobermorite to form.

BFS additions. With slag additions only, the hydrogarnite phase, siliated katoite (C$_3$ASH$_4$), is ultimately formed (Table 9.5) indicating a different formation mechanism to that of tobermorite (Figure 9.9). This was noted in previous work (Grant-Taylor *et al.*, 1996). This phase did not transform through to tobermorite, although

Table 9.4 Phases formed with PFA/quartz additions (XRD).

Composition	1 day	3 days	7 days	90 days
OPC:PFA (1:1)	Quartz (tr)	Amorphous and quartz (tr)	Tobermorite (tr) and amorphous	Tobermorite and amorphous
OPC:PFA (1:3)	Quartz (tr)	Quartz (tr)	Quartz (tr)	Amorphous
OPC:PFA:quartz (1:0.5: 0.5)	Quartz	Tobermorite (tr), quartz	Tobermorite, quartz	Tobermorite, quartz
OPC:PFA:quartz (1:1:1)	Quartz	Amorphous and quartz	Tobermorite (poorly crystalline), quartz	Tobermorite (poorly crystalline), quartz

OPC: ordinary Portland cement; PFA: pulverised fly ash; XRD: X-ray diffraction; (tr): trace.

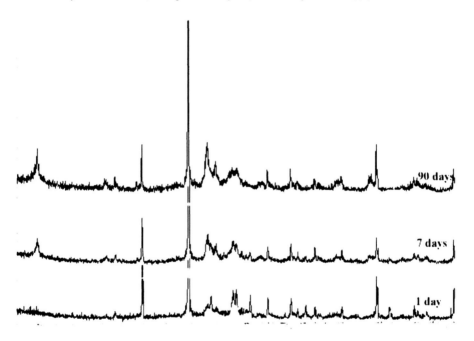

Figure 9.8 XRD traces of a 1:0.5:0.5 OPC:PFA:quartz blend autoclaved at 160°C. The basal peak for tobermorite at 7.6° 2θ is clearly visible after 7 days. OPC: ordinary Portland cement; PFA: pulverised fly ash; XRD: X-ray diffraction.

it did coexist with αC_2SH, indicating that the Ca:Si ratio has not been reduced far enough for the formation of tobermorite. The presence of $Ca(OH)_2$ confirms the high Ca:Si ratio. Gehlinite does not appear to react, but small amounts of hydrotalcite are formed from the magnesium in the slag. With increasing amounts of quartz, α-C_2SH_2 is formed until the Ca:Si ratio reduces to the level where tobermorite can form in small amounts by 90 days (Figure 9.10).

219

Table 9.5 Phases formed with BFS/quartz additions (XRD).

Composition	1 day	7 days	90 days
OPC:BFS (3:2)	CH + amorphous	CH, katoite amorphous	CH, katoite, α-C$_2$SH
OPC:BFS (4:3)	CH + amorphous	α-C$_2$SH (tr) amorphous	α-C$_2$SH katoite, a
OPC:BFS (3:4)	Amorphous	Katoite (tr), amorphous	α-C$_2$SH, katoite
Pure BFS	Amorphous	Amorphous	α-C$_2$SH (tr)
OPC:BFS:quartz (1:9:4)	Amorphous, quartz	Amorphous, quartz	α-C$_2$SH (tr), katoite
OPC:BFS:quartz (5:15:8)	Amorphous, quartz		Tobermorite (tr) quartz
BFS:quartz (5:2)	Amorphous + quartz	α-C$_2$SH (tr), quartz	Tobermorite (tr) quartz

OPC: ordinary Portland cement; BFS: blastfurnace slag; XRD: X-ray diffraction; (tr): trace.

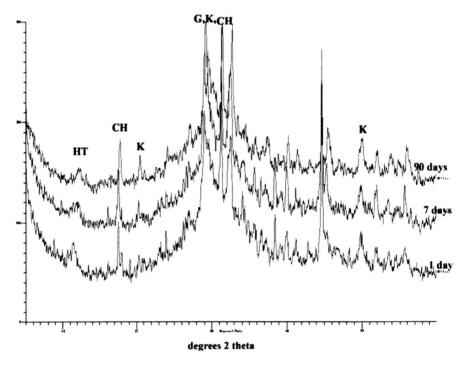

Figure 9.9 XRD traces of 1:3 OPC:BFS autoclaved at 160°C. K: katoite; HT: hydrotalcite; CH: Ca(OH)$_2$; G: gehlinite; OPC: ordinary Portland cement; BFS: blastfurnace slag.

9.6.2 Addition of Microsilica 600

Microsilica 600 (MS600) is an amorphous hydrothermal silica available in New Zealand (supplied by Golden Bay Microsilica). It has a typical particle size of 1–10 μm. The use of this material as a pozzolan for concrete was described by Chisholm (1997). It reacts quickly with Ca(OH)$_2$, unlike quartz, which requires hydrothermal conditions.

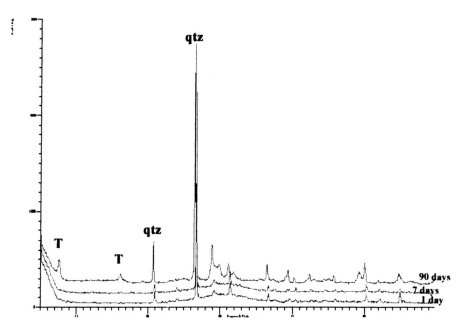

Figure 9.10 XRD trace of OPC:BFS:quartz =5:15:8 at 160°C. OPC: ordinary Portland cement; BFS: blastfurnace slag; XRD: X-ray diffraction.

Table 9.6 Compressive strengths of MS600 additions.

Silica in blend (%)	168 day UCS (MPa at 20°C)	84 day UCS (MPa at 90°C)	168 day UCS (MPa at 90°C)	28 day UCS (MPa at 150°C)	84 day UCS (MPa at 150°C)	168 day UCS (MPa at 150°C)	28 day UCS (MPa at 300°C)	84 day UCS (MPa at 300°C)
0	19	47	54	–	45	28	13	18
10	55	47	42	7.7	3	–	–	–
20	45	47	53	65	38	57	38	30

UCS: uniaxial compressive stress.

In recent work, Bigley *et al.* (2010) described additions of MS600 to geothermal well cements. With 20% addition and hydrothermal treatment, an amorphous binder is formed that has moderate strength, sufficient for well purposes, and is slow to carbonate and performs well up to 300°C.

To demonstrate its use, blends of Class G cement and 10, 15 and 20% additions MS600 were prepared, cast in 100 × 100 × 100 mm cubes and cured at 90°C before autoclaving at 150°C and 300°C for periods up to 6 months with an overpressure of CO_2 of 6 bar. Compressive strengths of various addition rates of MS600 are shown in Table 9.6. These strengths are less than those obtained with 40% quartz addition but adequate for well cementing. The binder is amorphous and remains so for up to

6 months curing at 300°C. Carbonation of this binder is limited to a thin carbonation sheath.

9.7 Discussion

The standard practice in geothermal cementing has been to follow the work of Kalousek (1955) by adding 40% or more of fine crystalline quartz to an API Class G or H cement designed for higher temperatures when cementing above 120°C. With time, this combination forms the low-Ca/Si-ratio phase of 11 Å tobermorite between 150 and 200°C, a strong impervious binder that converts to xonotlite at higher temperatures. The formation of these phases prevents the phenomenon known as *strength retrogression*. In commercial fibre cement boards where 40–60% quartz is added, only the finer quartz particles fully react after ~16 hours autoclaving at around 170°C, giving a rim of tobermorite around the coarser particles that acts as reinforcement. Klimesh (1999) showed that by adding an aluminous material to the slurry, the reaction of tobermorite formation was speeded up with the added advantage that it reduced 'water movement', the expansion and contraction of the board due to water uptake as a result of crosslinking of the silicate chains in the binder.

Carbonation of tobermorite (to aragonite at ~150°C) and of its higher-temperature counterpart xonotlite (to calcite) results in a loss of solid binder volume. Calcium migrates to the outer layers, leaving a leached zone (Milestone *et al.*, 1986a) and a discontinuity. A silica gel is formed and although the calcium carbonate contributes to strength, the composite becomes porous so carbonation continues (Milestone *et al.*, 1986c, 1987a). Eventually, carbonation will reach the centre of the sample. On the other hand, carbonation of the high-Ca/Si-ratio low-strength phases such as α-C_2SH (150–200°C) or reinhardbraunsite and kilchoanite (above 200°C) is an expansive reaction that forms an impervious carbonation sheath. This sheath contributes to strength and slows further attack. Depending on the slurry density this sheath varies in thickness from a few millimetres to around 10 mm in 6 months. In low-temperature CO_2 sequestration wells, this sheath is expected to last for many years (Kutchko *et al.*, 2008; Duguid, 2009).

The fluids in the New Zealand geothermal fields contain significant amounts of dissolved CO_2. Where geological conditions have created an impermeable cap-rock, CO_2 gas generated from the boiling zone at depth becomes trapped in an aquifer, usually at a lower temperature than the production fluid, so the dissolved CO_2 concentration is high at levels that are sufficient that the fluid is mildly acidic due to 'carbonic acid'. This acidic fluid is sufficient to both carbonate and dissolve any cement based on calcium silicates in what is known as a 'corrosive zone'. In some cases, samples exposed in this corrosive zone lost up to 60% of cement after only 6 months exposure, and even marble will start to dissolve.

Our experiments have also shown that if carbonation takes place at the same time as curing, this can upset the expected curing reactions by turning the $Ca(OH)_2$, as soon as it forms from cement hydration, into calcium carbonate (Milestone *et al.*, 2012). Thus, two competing reactions occur, carbonation and reaction of the $Ca(OH)_2$ with silica. If the silica is quartz, the reaction is slow and large amounts remain in a carbonated matrix. This carbonation reaction effectively removes available Ca, lowering the available Ca/Si ratio in the matrix, so small amounts of tobermorite are often seen where they were not expected with the Ca/Si ratio in the slurry. The temperature of initial hydration and curing is also critical. Flash setting can occur if the initial hydration takes place above 90°C and the expected curing reactions do not take place.

Why only quartz forms tobermorite is not clear, but it is thought to be due to the slow release of silica in an alkaline environment (Hong and Glasser, 2004: Jupe *et al.*, 2008). Other forms of silica do not form tobermorite, so their use has not been investigated fully, although Luke (2004) and Ma and Brown (1997) showed that if fly ash was mixed with quartz, or NaOH was added, then tobermorite would form. Milestone *et al.* (1992) showed that if only 20% silica flour was added, a new crystalline phase, killalaite with a Ca/Si ratio of 1.5, was formed, but more importantly it removed $Ca(OH)_2$ and so reduced the effects of strength retrogression. The phase was not as strong a binder as tobermorite, but when carbonated it formed an impermeable sheath slowing further carbonation. This addition was recommended as a compromise between strength and carbonation resistance for New Zealand geothermal wells.

Further work (Bigley *et al.*, 2010) has identified that the use of a very reactive silica, Microsilica 600, leads to an amorphous calcium silicate hydrate binder that shows little crystallisation, even when cured at 300°C for 6 months. The binder has moderate strength, which does not change, and carbonation is slow, with the formation of a carbonation sheath. Addition of this material at 25% BWOC with Class G cement is now the accepted practice for New Zealand geothermal well cementing. Microsilica 600 reacts rapidly with any $Ca(OH)_2$ formed, competing with the carbonation reaction, and lowers the pH in the slurry pores, probably preventing any crystallisation of calcium silicate hydrate phases. The same occurs with fly ash. Ma and Brown (1997) did find that if NaOH was added to a fly ash blend then tobermorite would form. The small amount of quartz in the PFA used in our experiments does ultimately allow tobermorite crystallisation in the same way as seen by Luke (2004), but the amount formed is small.

It would seem that the use of BFS might also be suitable in generating a carbonation-resistant phase. However, while the initial binder formed with BFS additions is amorphous, it ultimately crystallises to α-C_2SH and katoite, a very low-strength calcium aluminosilicate hydrate, C_3ASH_4, as seen in Section 9.6. Katoite was also found in the samples tested by Grant-Taylor *et al.* (1996, 1998). Although it is a very weak binder, it resists carbonation and is not suitable as an addition to cement for geothermal well applications.

9.8 Concluding remarks

The standard practice of adding ~40% silica flour to Portland well cements to provide relief from strength retrogression is unsuitable if the geothermal fluids contain significant amounts of dissolved CO_2. Any tobermorite formed carbonates rapidly and, if the CO_2 level is high enough, the carbonated layer dissolves rapidly or corrodes, leaving the steel casing unprotected.

The 20% addition of an amorphous hydrothermal silica, Microsilica 600, has been shown to be a very suitable compromise between strength and carbonation resistance. An impervious carbonation sheath forms and strength does not fall away with time.